Frederick Newton Willson

Practical Engineering Drawing and Third Angle Projection

For Students

Frederick Newton Willson

Practical Engineering Drawing and Third Angle Projection
For Students

ISBN/EAN: 9783744646468

Printed in Europe, USA, Canada, Australia, Japan

Cover: Foto ©berggeist007 / pixelio.de

More available books at **www.hansebooks.com**

PRACTICAL ENGINEERING DRAWING

AND

THIRD ANGLE PROJECTION

F. N. WILLSON

PRACTICAL ENGINEERING DRAWING
AND
THIRD ANGLE PROJECTION

FOR STUDENTS IN SCIENTIFIC, TECHNICAL, AND MANUAL TRAINING SCHOOLS

AND FOR

ENGINEERING AND ARCHITECTURAL DRAUGHTSMEN, SHEET METAL WORKERS, Etc.

BY

Frederick Newton Willson, C.E., A.M.,
Professor of Descriptive Geometry, Stereotomy and Technical Drawing
in the
John C. Green School of Science, Princeton University.

PUBLISHED BY THE AUTHOR
PRINCETON, N. J.
1907

ALTHOUGH the chapters here presented are taken from *Theoretical and Practical Graphics*—the author's more extended treatise on the theory and applications of descriptive geometry and mechanical drawing, they were prepared with a view to their separate issue in this form, and are independent of the other matter with which they are paged.

TABLE OF CONTENTS

NOTE-TAKING, DIMENSIONING, ETC.

Technical Free-Hand Sketching and Lettering. — Note-Taking from Measurement. — Dimensioning. — Conventional Representations.

Pages 5-10.

THE DRAUGHTSMAN'S EQUIPMENT.

The Choice and Use of Drawing Instruments and the Various Elements of the Draughtsman's Equipment. — General remarks preliminary to instrumental work.

Pages 11-20.

EXERCISES FOR PEN AND COMPASS.

Kinds and Signification of Lines. — Designs for Elementary Practice with the Right Line Pen. — Standard Methods of Representing Materials. — Line Shading. — Plane Problems of the Right Line and Circle, including Rankine's and Kochansky's approximations. — Exercises for the Compass and Bow-pen, including uniform and tapered curves. — The Anchor Ring. — The Hyperboloid. — A Standard Rail Section.

Pages 21-38.

ON HIGHER PLANE CURVES AND THE HELIX.

Regarding the Irregular Curve. — The Helix. — The Ellipse, Hyperbola and Parabola, by various methods of construction. — Homological Plane Curves. — Relief-Perspective. — Link-Motion Curves. — Centroids. — The Cycloid. — The Companion to the Cycloid. — The Curtate and Prolate Trochoids. — Hypo-, Epi-, and Peri-Trochoids. — Special Trochoids, as the Ellipse, Straight Line, Limaçon, Cardioid, Trisectrix, Involute and Spiral of Archimedes. — Parallel Curves. — Conchoid. — Quadratrix. — Cissoid. — Tractrix. — Witch of Agnesi. — Cartesian Ovals. — Cassian Ovals. — Catenary. — Logarithmic Spiral. — Hyperbolic Spiral. — Lituus. — Ionic Volute.

Pages 30-78.

TINTING AND SHADING.

Brush Tinting, Flat and Graduated. — Masonry, Tiling, Wood Graining, River-Beds, etc., with brush alone, or in combined brush and line work.

Pages 79-87.

THE LETTERING OF DRAWINGS.

Free-Hand Lettering. — Mechanical Expedients. — Proportioning of Titles. — Discussion of Forms. — Half-Block, Full Block and Railroad Types. — Borders and how to draw them. (Alphabets in Appendix).

Pages 88-96.

BLUE-PRINTING AND OTHER PROCESSES.

The Blue-print Process. — Photo-, and other Reproductive Graphic Processes, including Wood Engraving, Cerography, Lithography. Photo-lithography, Chromo-lithography, Photo-engraving, "Half-Tones," Photo-gravure and allied processes. — How to Prepare Drawings for Illustration.

Pages 97-103.

THIRD ANGLE PROJECTION. — WORKING DRAWINGS.

Projections and Intersections by the Third Angle Method. — The Development of Surfaces, for Sheet Metal or Arch Constructions. Working Drawings of Bridge Post Connection. — Structural Iron. — Spur Gearing (Approximate Involute Outlines). — Helical Springs, Rectangular and Circular Section. — Screws and Bolts (U. S. Standard), and Table of Proportions.

Pages 131-180.

AXONOMETRIC (INCLUDING ISOMETRIC) PROJECTION. — ONE-PLANE DESCRIPTIVE GEOMETRY.

Orthographic Projection upon a Single Plane. — Axonometric Projection. — General Fundamental Problem, inclinations known for two of the three axes. Isometric Projection vs. Isometric Drawing. — Shadows on Isometric Drawings. — Timber Framings and Arch Voussoirs in Isometric View. — One-Plane Descriptive Geometry.

Pages 241-247.

OBLIQUE PROJECTION.

Oblique or Clinographic Projection, Cavalier Perspective, Cabinet Projection, Military Perspective. — Applications to Timber Framings, Arch Voussoirs and Drawing of Crystals.

Pages 248-250.

APPENDIX.

Table of the Proportions of Washers. — Working Drawings of Standard 100-lb. Rail, and of Allen-Richardson Slide Valve. — Designs for Variation of Problems in Chapters X, XV and XVI. — Alphabets.

Pages 251-268.

CHAPTER II.

ARTISTIC AND TECHNICAL FREE-HAND DRAWING.—SKETCHING FROM MEASUREMENT.—FREE-HAND LETTERING.—CONVENTIONAL REPRESENTATIONS.

20. Drawings, if classified as to the *method of their production*, are either *free-hand* or *mechanical*; while as to *purpose* they may be *working drawings*, so fully dimensioned that they can be worked from and what they represent may be manufactured; or *finished* drawings, illustrative or artistic in character and therefore shaded either with pen or brush, and having no hidden parts indicated by dotted lines as in the preceding division. Finished drawings also lack figured dimensions.

Working drawings of parts or "details" of a structure are called *detail drawings;* while the representation of a structure as a whole, with all its details in their proper relative position, hidden parts indicated by dotted lines, etc., is termed a *general* or *assembly* drawing.

21. While mechanical drawing is involved in making the various essential views—plans, elevations and sections—of all engineering and architectural constructions, and in solving the problems of *form* and *relative position* arising in their design, yet, to the engineer, the ability to sketch effectively and rapidly, *free-hand*, is of scarcely less importance than to handle the drawing instruments skillfully; while the success of an architect depends in still greater measure upon it.

We must distinguish, however, between *artistic* and *technical* free-hand work. The architect must be master of both; the engineer necessarily only of the latter.

To secure the adoption of his designs the architect relies largely upon the effective way in which he can finish, either with pen and ink or in water-colors, the perspectives of exterior and interior views; and such drawings are judged mainly from the artistic standpoint. While it is not the province of this treatise to instruct in such work a word of suggestion may properly be introduced for the student looking forward to architecture as a profession. He should procure Linfoot's *Picture Making in Pen and Ink*, Miller's *Essentials of Perspective* and Delamotte's *Art of Sketching from Nature;* and with an experienced architect or artist, if possible, but otherwise by himself, master the principles and act on the instructions of these writers.

22. Since the camera makes it, fortunately, no longer essential that a civil engineer should be a landscape artist as well, his free-hand work has become more restricted in its scope and more rigid in its character, and like that of the machine designer it may properly be called *technical*, from its object. Yet to attain a sufficient degree of skill in it for all practical and commercial purposes is possible to all, and among them many who could never hope to produce artistic results. It is confined mainly to the making of *working sketches*, *conventional representations* and *free-hand lettering*, and the equipment therefor consists of a pencil of medium grade as to hardness; lettering pens—Falcon or Gillott's 303, with Miller Bros. "Carbon" pen No. 4; either a note-book or a sketch-block or pad; also the following for sketching from measurement: a two-foot pocket-rule; calipers, both external and internal, for taking outside and inside diameters; a pair of pencil compasses for making an occasional circle too large to be drawn absolutely free-hand; and a steel tape-measure for large work, if one can have assistance in taking notes, but otherwise a long rod graduated to eighths.

6   *THEORETICAL AND PRACTICAL GRAPHICS.*

23. In the evolution of a machine or other engineering project the designer places his ideas on paper in the form of rough and mainly free-hand sketches, beginning with a general outline, or "skeleton" drawing of the whole, on as large a scale as possible, then filling in the details, separate —and larger—drawings of which are later made to exact scale. While such preliminary sketches are not drawn literally "to scale" it is obviously desirable that something like the relative proportions should be preserved and that the closer the approximation thereto the clearer the idea they will give to the draughtsman or workman who has to work from them. A habit of close observation must therefore be cultivated, of analysis of form and of relative direction and proportion, by all who would succeed in draughting, whether as designers or merely as copyists of existing constructions. While the beginner belongs necessarily in the latter category he must not forget that his aim should be to place himself in the ranks of the former, both by a thorough mastery of the fundamental theory that lies back of all correct design and by such training of the hand as shall facilitate the graphic expression of his ideas. To that end he should improve every opportunity to put in practice the following instructions as to

SKETCHING FROM MEASUREMENT,

as each structure sketched and measured will not only give exercise to the hand but also prove a valuable object lesson in the proportioning of parts and the modes of their assemblage.

A free-hand sketch may be as good a working drawing as the exactly scaled—and usually *inked*—drawing that is generally made from it to be sent to the shop.

While several views are usually required, yet for objects of not too complicated form, and whose lines lie mainly in mutually perpendicular directions, the method of representation illustrated by Fig. 7, is admirably adapted,* and obviates all necessity for additional sketches. It is an *oblique projection*

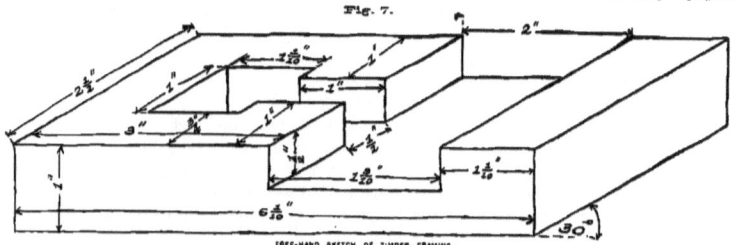

Fig. 7.

FREE-HAND SKETCH OF TIMBER FRAMING.

(Art. 17) the theory of whose construction will be found in a subsequent chapter, but with regard to which it is sufficient at this point to say that the right angles of the front face are seen in their true form, while the other right angles are shown either of 30°, 60°, or 120°; although almost any oblique angle will give the same general effect and may be adopted. Lines parallel to each other on the object are also parallel in the drawing.

Draw first the front face, whose angles are seen in their true form; then run the oblique lines off in the direction which will give the best view. (Refer to Figs. 42, 44, 45 and 46.)

24. While Fig. 7 gives almost the pictorial effect of a true perspective and the object requires no other description, yet for complicated and irregular forms it gives place to the plan-and-elevation mode of representation, the plan being a *top* and the elevation a *front* view of the object. And

---

* The figures in this chapter are photo-reproductions of free-hand work and are intended not only to illustrate the text but also to set a reasonable standard for sketch-notes.

if two views are not enough for clearness as many more should be added as seem necessary, including what are called *sections*, which represent the object as if cut apart by a plane, separated and a view obtained perpendicular to the cutting plane, showing the internal arrangement and shape of parts.

In Fig. 8 we have the same object as in Fig. 7, but represented by the method just mentioned. The front view (elevation) is evidently the same in both Figs. 7 and 8, except that in the latter we indicate by dotted lines the hidden recess which is in full sight in Fig. 7.

The view of the top is placed *at the top* in conformity to the now quite general practice as to location, viz., grouping the various sketches about the elevation, so that the view of the left end is *at the left*, of the right *at the right*, etc.

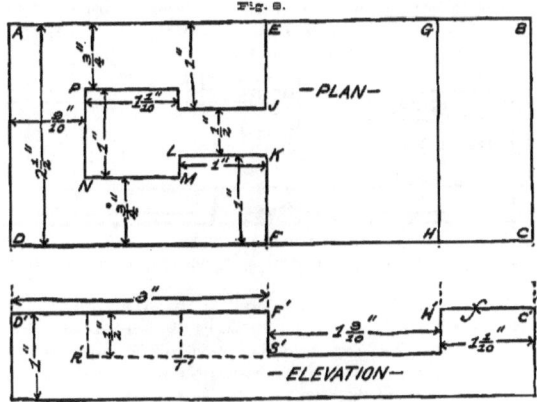

Fig. 8.

FREE-HAND SKETCH OF TIMBER FRAMING, IN PLAN AND ELEVATION.

In these views, which fall under Art. 19 as to theoretical construction, entire surfaces are projected as straight lines, as $GBCH$ in the straight line $H'C'$. Were this a metallic surface and "*finished*" or "*machined*" to smoothness, as distinguished from the surface of a rough casting, that fact would be denoted by an "*f*" on the line $H'C'$ which represents the entire surface, the cross-line of the "*f*" cutting the line obliquely, as shown.

CENTRE-LINES.—DIMENSIONING.

25. *Dimensioning.* In sketching, centre-lines and all important centres should be located first, and measurements taken from them or from finished surfaces.

Feet and inches are abbreviated to "Ft.," and "In.," as 4 Ft. $6\frac{3}{8}$ In.; also written 4′ $6\frac{3}{8}″$, and occasionally 4 Ft. $6\frac{3}{8}″$. A dimension should not be written as an improper fraction, $\frac{13}{8}″$ for example, but as a mixed number, $1\frac{5}{8}″$. Fractions should have *horizontal* dividing lines.

Not only should dimensions of successive parts be given but an "over-all" dimension, which, it need hardly be said, should sustain the axiom regarding the whole and the sum of its parts.

Dimensions should read in line with the line they are on, and either from the bottom or the right hand.

The arrow tips should touch the lines between which a distance is given.

8   *THEORETICAL AND PRACTICAL GRAPHICS.*

Extension lines should be drawn and the dimension given *outside* the drawing whenever such course will add to the clearness. (See $D'F'$, Fig. 8.)

An opening should always be left in the dimension line for the figures.

In case of very small dimensions the arrow tips may be located outside the lines, as in Fig. 9, and the dimension indicated by an arrow, as at A, or inserted as at B if there is room.

Should a piece of *uniform cross-section* (as, for example, a rail, angle-iron, channel bar, Phœnix column or other form of structural iron) be too long to be represented in its proper relative length on the sketch it may be broken as in Fig. 9, and the *form* of the section (which in the case supposed will be the same as an *end view*) may be inserted with its dimensions, as in the shaded figure. If the kind of bar and the number of pounds per yard are known the dimensions can be obtained by reference to the handbook issued by the manufacturers.

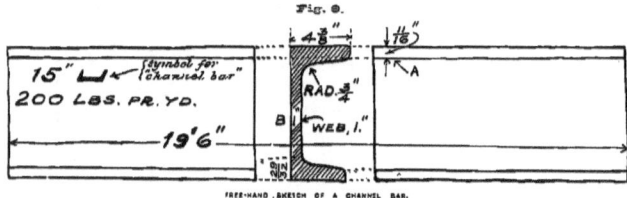

FREE-HAND SKETCH OF A CHANNEL BAR.

The same dimension should not appear on each view, but each dimension must be given at least once on some view.

Notes on Riveted Work, Pins, Bolts, Screws and Nuts. In riveted work the "pitch" of the rivets, i. e., their distance from centre to centre ("c. to c.") should be noted, as also that between centre lines or rows, and of the latter from main centre lines. Similarly for bolts and holes. If the latter are located in a circle note the diameter of the circle containing their centres. Note that a hole for a rivet is usually about one-half the diameter of the forged head.

In measuring nuts take the width between parallel sides ("width across the flats") and abbreviate for the shape, as "sq.," "hex.," "oct."

For a piece of cylindrical shape a frequently used symbol is the circle, as 4" ○ (read "four inches, round," *not around,*) for 4" diameter; but it is even clearer to use the abbreviation of the latter word, viz., "diam."

In taking notes on bolts and screws the outside diameter is sufficient if they are "standard," that is, proportioned after either the Sellers (U. S. Standard) or Whitworth (English Standard) systems, as the proportions of heads and nuts, number of threads to the inch, etc., can be obtained from the tables in the Appendix. If not "standard" note the number of threads to the inch. Record whether a screw is right- or left-handed. If right-handed it will advance if turned clock-wise. The shape of thread, whether triangular or square, would also be noted.

Notes on Gearing. On *cog*, or "gear," wheels obtain the distance between centres and the number of teeth on each wheel. The remaining data are then obtained by calculation.

Bridge Notes. In taking bridge notes there would be required general sketches of front and end view; of the flooring system, showing arrangement of tracks, ties, guard-beam and side-walk; a cross-section; also detail drawings of the top and foot of each post-connection in one longitudinal line from one end to the middle of the structure. In case of a double-track bridge the outside rows of posts are alike but differ from those of the middle truss.

CONVENTIONAL REPRESENTATIONS.—FREE-HAND LETTERING.   9

All notes should be taken on as large a scale as possible, and so indexed that drawings of parts may readily be understood in their relation to the whole.

The foregoing hints might be considerably extended to embrace other and special cases, but experience will prove a sufficient teacher if the student will act on the suggestions given, and will remember that to get an excess of data is to err on the side of safety. It need hardly be added that what has preceded is intended to be merely a partial summary of the instructions which would be given in the more or less brief practice in technical sketching which, presumably, constitutes a part of every course in Graphics; and that unless the draughtsman can be under the direction of a teacher he will be able to sketch much more intelligently after studying more of the theory involved in Mechanical Drawing and given in the later pages of this work.

CONVENTIONAL REPRESENTATIONS.

26. Conventional representations of the natural features of the country or of the materials of construction are so called on the assumption, none too well founded, that the engineering profession

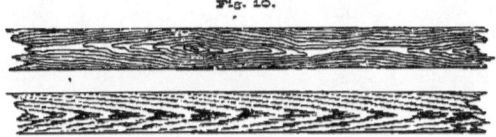
Fig. 10.

has agreed in convention that they shall indicate that which they also more or less resemble. While there is no universal agreement in this matter there is usually but little ambiguity in their use, especially in those that are drawn free-hand, since in them there can be a nearer approach to the natural appearance. This is well illustrated by Figs. 10 and 11.

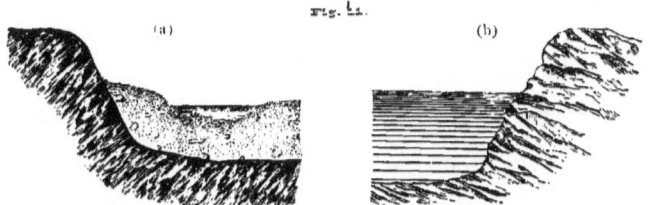
Fig. 11.

In addition to a rock section Fig. 11 (a) shows the method of indicating a mud or sand bed with small random boulders.

Water either in section or as a receding surface may be shown by parallel lines, the spaces between them increasing gradually.

Conventional representations of wood, masonry and the metals will be found in Chapter VI, after hints on coloring have been given, the foregoing figures appearing at this point merely to illustrate, in black and white, one of the important divisions of technical free-hand work. Those, however, who have already had some practice in drawing may undertake them either with pen and ink or in colors, in the latter case observing the instructions of Arts. 237–241 for wood, while for the river sections they may employ *burnt umber* undertone for the earthy bed, *pale blue* or *india ink* tint for the rock, and *prussian blue* for the water lines.

FREE-HAND LETTERING.

27. Although later on in this work an entire chapter is devoted to the subject of lettering, yet at this point a word should be said regarding those forms of letters which ought to be mastered, early in a draughting course, as the most serviceable to the practical worker.

Fig. 12.

A B C D E F G H I J K L M N O P Q R S T U V W X Y Z &
1 2 3 4 5 6 7 8 9 0

*A B C D E F G H I J K L M N O P Q R S T U V W X Y Z &
1 2 3 4 5 6 7 8 9 0*

The first, known as the *Gothic*, is the simplest form of letter, and is illustrated in both its vertical and inclined (or *Italic*) forms in Fig. 12. It is much used in dimensioning, as well as for titles. The lettering and numerals are Gothic in Figs. 7 and 8, with the exception of the 1 and 4, which, by the addition of feet, are no longer a pure form although enhanced in appearance.

Fig. 12 (a).

935 4

In Fig. 12 (a) some modifications of the forms of certain numerals are shown; also the omission of the dividing line in a mixed number, as is customary in some offices.

For Gothic forms a coarse pen is necessary, and shading is to be avoided, the object being to get all lines of uniform weight. Miller Bros'. "Carbon" pen, No. 4, is well adapted to them for work on a fairly smooth surface and with a free-flowing ink.

Fig. 13 illustrates the Italic (or *inclined*) form of a letter which when vertical is known as the *Roman*. The *Roman* and *Italic Roman* are much used on Government and other map work, and in

Fig. 13.

*A B C D E F G H I J K L M N O P Q R S
1 2 3 4 5   T U V W X Y Z   6 7 8 9 0
a b c d e f g h i j k l m n o p q r s t u v w x y z*

the draughting offices of many prominent mechanical engineers. Regarding them the student may profitably read Arts. 260–262. Make the spaces between letters as nearly uniform as possible, and the small letters usually about three-fifths the height of the capitals in the same line.

For Roman and other forms of letter requiring shading use a fine pen; Gillott's No. 303 for small work, and a "Falcon" pen for larger.

A form of letter much used in Europe and growing in favor here is the *Soennecken Round Writing*, referred to more particularly in Art. 265 and illustrated by a complete alphabet in the Appendix. The text-book and special pens required for it can be ordered through any dealer in draughtsmen's supplies.

CHAPTER III.

DRAWING INSTRUMENTS AND MATERIALS.—INSTRUCTIONS AS TO USE.—GENERAL PRELIMINARIES AND TECHNICALITIES.

28. The draughtsman's equipment for graphical work should be the best consistent with his means. It is mistaken economy to buy inferior instruments. The best obtainable will be found in the end to have been the cheapest.

The set of instruments illustrated in the following figures contains only those which may be considered absolutely essential for the beginner.

Fig. 14. Fig. 15.

THE DRAWING PEN.

The right line pen (Fig. 14) is ordinarily used for drawing straight lines, with either a rule or triangle to guide it; but it is also employed for the drawing of curves when directed in its motion by curves of wood or hard rubber. For average work a pen about five inches long is best.

The figure illustrates the most approved type, i. e., made from a single piece of steel. The distance between its points, or "nibs," is adjustable by means of the screw H. An older form of pen has the outer blade connected with the inner by a hinge. The convenience with which such a pen may be cleaned is more than offset by the certainty that it will not do satisfactory work after the joint has become in the slightest degree loose and inaccurate through wear.

29. If the points wear unequally or become blunt the draughtsman may sharpen them readily himself upon a fine oil-stone. The process is as follows:

Screw up the blades till they nearly touch. Incline the pen at a small angle to the surface of the stone and draw it lightly from left to right (supposing

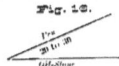

Fig. 16.

the initial position as in Fig. 16). Before reaching the right end of the stone begin turning the pen in a plane perpendicular to the surface, and draw in the opposite direction at the same angle. After frequent examination and trial, to see that the blades have become equal in length and similarly rounded, the process is completed by lightly dressing the outside of each blade separately upon the stone. No grinding should be done on the inside of the blade. Any "burr" or rough edge resulting from the operation may be removed with fine emery paper. For the best results, obtained in the shortest possible time, a magnifying glass should be used. The student should take particular notice of the shape of the pen when new, as a standard to be aimed at when compelled to act on the above suggestions.

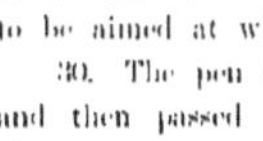

30. The pen may be supplied with ink by means of an ordinary writing pen dipped in the ink and then passed between the blades; or by using in the same manner a strip of Bristol board about a quarter of an inch in width. Should any fresh ink get on the outside of the pen it must

12   THEORETICAL AND PRACTICAL GRAPHICS.

be removed; otherwise it will be transferred to the edge of the rule and thence to the paper, causing a blot.

31. As with the pencil, so with the pen, horizontal lines are to be drawn *from left to right*, while vertical or inclined lines are drawn either from or toward the worker, according to the position of the guiding edge with respect to the line to be drawn. If the line were *m n*, Fig. 17, the motion would be away from the draughtsman, i. e., from *n* toward *m*; while *o p* would be drawn *toward* the worker, being on the right of the triangle.

Fig. 17.

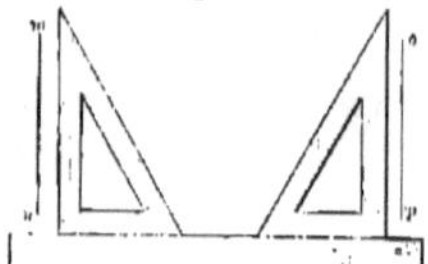

32. To make a sharply defined, clean-cut line—the only kind allowable—the pen should be held lightly but firmly with one blade resting against the guiding edge, and with both points resting equally upon the paper so that they may wear at the same rate.

33. The inclination of the pen to the paper may best be about 70°. When properly held the pen will make a line about a fortieth of an inch from the edge of the rule or triangle, leaving visible a white line of the paper of that width. If, then, we wish to connect two points by an inked straight line, the rule must be so placed that its edge will be from them the distance indicated.

It need hardly be said that a *drawing*-pen should not be *pushed*.

The more frequently the draughtsman will take the trouble to clean out the point of the pen and supply fresh ink the more satisfactory results will he obtain. When through with the pen clean it carefully, and lay it away with the points not in contact. Equal care should be taken of all the instruments, and for cleaning them nothing is superior to chamois skin.

DIVIDERS.

34. The hair-spring dividers (Fig. 15) are employed in dividing lines and spacing off distances, and are capable of the most delicate adjustment by means of the screw *G* and spring in one of the legs. When but one pair of dividers is purchased the kind illustrated should have the preference over plain dividers, which lack the spring. It will, however, be frequently found convenient to have at hand a pair of each. Should the joint at *F* become loose through wear it can be tightened by means of a key having two projections which fit into the holes shown in the joint.

35. In spacing off distances the pressure exerted should be the slightest consistent with the location of a point, the puncture to be merely in the surface of the paper and the points determined by lightly pencilled circles about them, thus ─◯───◯─── . In laying off several equal distances along a line all the arcs described by one **Fig. 18.** leg of the dividers should be on the same side of the line. Thus, in Fig. 19, with *b* the first centre of turning, the leg *x* describes the

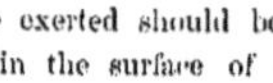

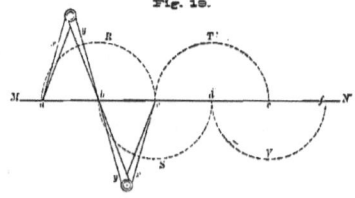

Fig. 19.

arc *R*, then rests and pivots on *c* while the leg *y* describes the arc *S*; *x* then traces arc *T*, etc.

THE COMPASSES.—BOW-PENCIL AND PEN.

COMPASS SET.
Fig. 20. Fig. 21. Fig. 22.

36. The compasses (Fig. 20) resemble the dividers in form and may be used to perform the same office, but are usually employed for the drawing of circles. Unlike the dividers one or both of the legs of compasses are detachable. Those illustrated have one permanent leg, with pivot or "needle-point" adjustable by means of screw R. The other leg is detachable by turning the screw O, when the pen leg LM (Fig. 21) may be inserted for ink work; or, where large work is involved, the lengthening bar on the right (Fig. 22) may be first attached at O and the pencil or pen leg then inserted at l. The metallic point held by screw S is usually replaced by a hard lead, sharpened as indicated in Art 34.

37. When in use the legs should be bent at the joints P and L, so that they will be perpendicular to the paper when the compasses are held in a vertical plane. The turning may be in either direction, but is usually "clock-wise;" and the compasses may be slightly inclined toward the direction of turning. When so used, and if no undue pressure be exerted on the pivot leg, there should be but the slightest puncture at the centre, while the pen points having rested equally upon the paper have sustained equal wear, and the resulting line has been sharply defined on both sides. Obviously the legs must be re-adjusted as to angle, for any material change in the size of the circles wanted.

The compasses should be held and turned by the milled head which projects above the joint N.

Dividers and compasses should open and shut with an absolutely uniform motion and somewhat stiffly.

BOW-PENCIL AND PEN.

Fig. 23. Fig. 24.

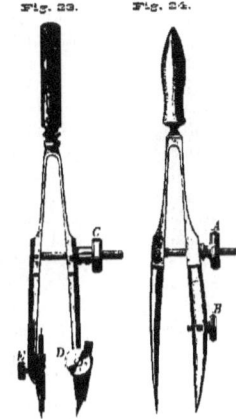

38. For extremely accurate work, in diameters from one-sixteenth of an inch to about two inches, the bow-pencil (Fig. 23) and bow-pen (Fig. 24) are especially adapted. The pencil-bow has a needle-point, adjustable by means of screw E, which gives it a great advantage over the fixed pivot-point of the bow-pen, not alone in that it permits of more delicate adjustment for unusually small work but also because it can be easily replaced by a new one in case of damage; whereas an injury to the other renders the whole instrument useless. For very small circles the needle-point should project very slightly beyond the pen-point; theoretically by only the extremely small distance the needle-point is expected to sink into the paper.

The spring of either bow should be strong; otherwise an attempt at a circle will result in a spiral.

It will save wear upon the threads of the milled heads A and C if the draughtsman will press the legs of the bow together with his left hand and run the head up loosely on the screw with his right.

39. To the above described—which we may call the *minimum* set of instruments—might be advantageously added a pair of bow-spacers (small dividers shaped like Fig. 24); beam-compasses, for extra large circles; parallel-rule; proportional dividers, and an extra—and larger—right-line pen.

40. The remainder of the *necessary* equipment consists of paper; a drawing-board; T-rule; triangles or "set squares;" scales; pencils; India ink; water colors; saucers for mixing ink or colors; brushes; water-glass and sponge; irregular (or "French") curves; india rubber; erasing knife; protractor; file for sharpening pencils, or a pad of fine emery or sand paper; thumb-tacks (or "drawing-pins"); horn centre, for making a large number of concentric circles.

PAPER AND TRACING CLOTH.

41. Drawing paper may be purchased by the sheet or roll and either unmounted or mounted, i. e., "backed" by muslin or heavy card-board. Smooth or "hot-pressed" paper is best for drawings in line-work only; but the rougher surfaced, or "cold-pressed," should always be employed when brush-work in ink or colors is involved: in the latter case, also, either mounted paper should be used or the sheets "stretched" by the process described in Art. 44.

42. The names and sizes of sheets are:—

Cap 13 × 17	Elephant 23 × 28
Demi 15 × 20	Atlas 26 × 34
Medium 17 × 22	Columbia 23 × 35
Royal 19 × 24	Double Elephant 27 × 40
Super Royal 19 × 27	Antiquarian 31 × 53

Imperial 22 × 30

43. There are many makes of first-class papers, but the best known and still probably the most used is Whatman's. The draughtsman's choice of paper must, however, be determined largely by the value of the drawing to be made upon it, and by the probable usage to which it will be subjected.

Where several copies of one drawing were desired it has been a general practice to make the original, or "construction" drawing, with the *pencil*, on paper of medium grade, then to lay over it a sheet of *tracing-cloth* and copy upon it, in ink, the lines underneath. Upon placing the tracing cloth over a sheet of sensitized paper, exposing both to the light and then immersing the sensitive paper in water, a copy or print of the drawing was found upon the sheet, in white lines on a blue ground—the well-known *blue-print*. The time of the draughtsman may, however, be economized, as also his purse, by making the original drawing in ink upon Crane's *Bond* paper, which combines in a remarkable degree the qualities of transparency and toughness. About as clear blue-prints can be made with it as with tracing-cloth, yet it will stand severe usage in the shop or the drafting-room.

Better papers may yet be manufactured for such purposes, and the progressive draughtsman will be on the alert to avail himself of these as of all genuine improvements upon the materials and instruments before employed.

44. To stretch paper tightly upon the board lay the sheet right side up,[*] place the long rule with its edge about one-half inch back from each edge of the paper in turn, and fold up against it a margin of that width. Then *thoroughly* dampen the *back* of the paper with a full sponge, except on the folded margins. Turning the paper again face up gum the margins with strong mucilage or glue, and quickly but firmly press *opposite* edges down simultaneously, long sides first, exerting at the same time a slight outward pressure with the hands to bring the paper down somewhat closer to

---

[*] The "right side" of a sheet is, presumably, that toward one when—on holding it up to the light—the manufacturer's name, in water-mark, reads correctly.

the board. Until the gum "sets," so that the paper adheres perfectly where it should, the latter should not shrink; hence the necessity for so completely soaking it at first. The sponge may be applied to the *face* of the paper provided it is not *rubbed* over the surface, so as to damage it. The stretch should be horizontal when drying, and no excess of water should be left standing on the surface; otherwise a water-mark will form at the edge of each pool.

45. When tracing-*cloth* is used it must be fastened smoothly, with thumb-tacks, over the drawing to be copied, and the ink lining done upon the glazed side, any brush work that may be required — either in ink or colors — being always done upon the dull side of the cloth after the outlining has been completed.

If the glazed surface be first dusted with powdered pipe-clay applied with chamois skin it will take the ink much more readily.

When erasure is necessary use the rubber, after which the surface may be restored for further pen-work by rubbing it with soapstone.

Tracing-cloth, like drawing paper, is most convenient to work upon if perfectly flat. When either has been purchased by the roll it should therefore be cut in sheets and laid away for some time in drawers to become flat before needed for use.

DRAWING BOARD.

46. The drawing board should be slightly larger than the paper for which it is designed and of the most thoroughly seasoned material, preferably some *soft* wood, as pine, to facilitate the use of the drawing-pins or thumb-tacks. To prevent warping it should have battens of hard wood dovetailed into it across the back, transversely to its length. The back of the board should be grooved longitudinally to a depth equal to half the thickness of the wood, which weakens the board transversely and to that degree facilitates the stiffening action of the battens.

For work of moderate size, on stretched paper, yet without the use of mucilage, the "panel" board is recommended, provided that both frame and panel are made of the best seasoned hard wood.

It will be found convenient for each student in a technical school to possess two boards, one 20" × 28" for paper of Super Royal size, which is suitable for much of a beginner's work, and another 28" × 41" for Double Elephant sheets (about twice Super Royal size) which are well adapted to large drawings of machinery, bridges, etc. A large board may of course be used for small sheets, and the expense of getting a second board avoided; but it is often a great convenience to have a medium-sized board, especially in case the student desires to do some work outside the draughting-room.

THE T-RULE.

47. The T-rule should be slightly shorter than the drawing board. Its head and blade must have absolutely straight edges, and be so rigidly combined as to admit of no lateral play of the latter in the former. The head should also be so fastened to the blade as to be level with the surface of the board. This permits the triangles to slide freely over the head, a great convenience when the lines of the drawing run close to the edge of the paper. (See Fig. 32.)

The head of the T-rule should always be used along the left-hand edge of the drawing board.

TRIANGLES.

48. Triangles, or "set-squares" as they are also called, can be obtained in various materials, as hard rubber, celluloid, pear-wood, mahogany and steel; and either solid (Fig. 25) or open (Fig. 26). The open triangles are preferable, and two are required, one with acute angles of 30° and 60°, the other with 45° angles. Hard rubber has an advantage over metal or wood, the latter being likely to warp and the former to rust, unless plated. Celluloid is transparent and the most cleanly of all.

16   *THEORETICAL AND PRACTICAL GRAPHICS.*

The most frequently recurring problems involving the use of the triangles are the following:—

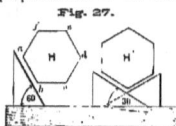

49. *To draw parallel lines* place either of the edges against another triangle or the T-rule. If then moved along, in either direction, each of the other edges will take a series of parallel positions.

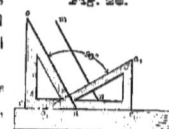

50. *To draw a line perpendicular to a given line* place the *hypotenuse* of the triangle, $o\,a$, (Fig. 26), so as to coincide with or be parallel to the given line; then a rule or another triangle against the base. By then turning the triangle so that the other side, $o\,c$, of its right angle shall be against the rule, as at $o_1 c_1$, the hypotenuse will be found perpendicular to its first position and therefore to the given line.

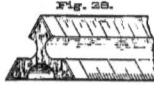

51. *To construct regular hexagons* place the *shortest* side of the 60° triangle against the rule (Fig. 27) if two sides are to be *horizontal*, as fe and $b\,c$ of hexagon H. For vertical sides, as in H', the position of the triangle is evident. By making $a\,b$ indefinite at first, and knowing $b\,c$ — the length of a side, we may obtain a by an arc, centre b, radius $b\,c$.

If the inscribed circles were given, the hexagons might also be obtained by drawing a series of tangents to the circles, with the rule and triangles in the positions indicated.

THE SCALE.

52. But rarely can a drawing be made of the same size as the object, or "full-size," as it is called; the lines of the drawing, therefore, usually bear a certain ratio to those of the object. This ratio is called the *scale* and should invariably be indicated.

If six inches on the drawing represent one foot on the object the scale is *one-half* and might be variously indicated, thus: SCALE $\frac{1}{2}$; SCALE 1:2; SCALE 6 IN. = 1 FT. SCALE 6" = 1'.

At one foot to the inch any line of the drawing would be one-twelfth the actual size, and the fact indicated in either of the ways just illustrated.

Although it is a simple matter for the draughtsman to make a scale for himself for any particular case yet scales can be purchased in great variety, the most serviceable of which for the usual range of work is of box-wood, 12" long, (or 18", if for large work) of the form illustrated by Fig. 28, and graduated $\frac{3}{32}:\frac{3}{16}:\frac{1}{8}:1:\frac{3}{8}:\frac{1}{2}:\frac{3}{4}:1:1\frac{1}{2}:3$ inches to the foot. This is known as the *architect's* scale in contradistinction to the *engineer's*, which is decimally graduated. It will, however, be frequently convenient to have at hand the latter as well as the former.

When in use it should be laid along the line to be spaced, and a light dot made upon the

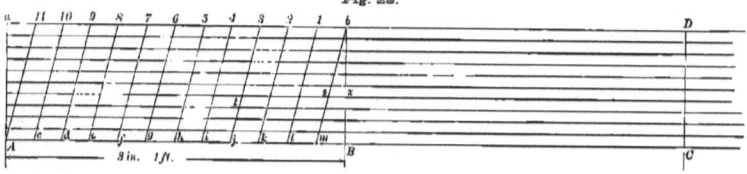

paper with the pencil, opposite the proper division on the graduated edge. A distance should rarely be transferred from the scale to the drawing by the dividers, as such procedure damages the scale if not the paper.

53. For special cases *diagonal* scales can readily be constructed. If, for example, a scale of 3 inches to the foot is needed and measuring to *fortieths of inches*, draw eleven equidistant, parallel lines, enclosing ten equal spaces, as in Fig. 29, and from the end A lay off AB, BC, etc., each 3 inches and representing a foot. Then twelve parallel diagonal lines in the first space intercept quarter-inch spaces on AB or ab, each representative of an inch. There being ten equal spaces between B and b, the distance sx, of the diagonal bm from the vertical bB, taken on any horizontal line sx, is as many tenths of the space mB as there are spaces between sx and b; six, in this case. The principle of construction may be generalized as follows: —

The distance apart of the vertical lines represents the units of the scale, whether inches, feet, rods or miles. Except for decimal graduation divide the left-hand space at top and bottom into as many spaces as there are units of the next lower denomination in one of the original units (feet, for yards as units; inches in case of feet, etc.). Join the points of division by diagonal lines; and, if $\frac{1}{x}$ is the smallest fraction that the scale is designed to give, rule $x+1$ equidistant horizontal lines, giving x equal horizontal spaces. The scale will then read to $\frac{1}{x}$th of the intermediate denomination of the scale.

When a scale is properly used, the spaces on it which represent feet and inches are treated as if they were such in fact. On a scale of one-eighth actual size the edge graduated 1½ inches to the foot would be employed; each 1½ inch space on the scale would be read as if it were a foot; and ten inches, for example, would be ten of the eighth-inch spaces, each of which is to represent an inch of the original line being scaled. The usual error of beginners would be to divide each original dimension by eight and lay off the result, actual size. The former method is the more expeditious.

THE PENCILS.

54. For construction lines afterward to be inked the pencils should be of *hard* lead, grade 6H if Faber's or VVH if Dixon's. The pencilling should be *light*. It is easy to make a groove in the paper by exerting too great pressure when using a hard lead. The hexagonal form of pencil is usually indicative of the finest quality, and has an advantage over the cylindrical in not rolling off when on a board that is slightly inclined.

Somewhat softer pencils should be used for drawings afterward to be traced, and for the preliminary free-hand sketches from which exact drawings are to be made; also in free-hand lettering.

Sharpen to a *chisel* edge for work along the edges of the T-rule or triangles, but use another pencil with a *coned* point for marking off distances with a scale, locating centres and other isolated points, and for free-hand lettering; also sharpen the compass leads to a point. Use the knife for cutting the wood of the pencil, beginning at least an inch from the end. Leave the lead exposed for a quarter of an inch and shape it as desired, either with a knife or on a fine file, or a pad of emery paper.

THE INK.

55. Although for many purposes some of the liquid drawing-inks now in the market, particularly Higgins', answer admirably, yet for the best results, either with pen or brush, the draughtsman should mix the ink himself with a stick of India — or, more correctly, *China* ink, selecting one of the higher-priced cakes, of rectangular cross-section. The best will show a lustrous, almost iridescent fracture, and will have a smooth, as contrasted with a gritty *feel* when tested by rubbing the moistened finger on the end of the cake.

Sets of saucers, called "nests," designed for the mixing of ink and colors, form an essential part of an equipment. There are usually six in a set and so made that each answers as a cover for the one below it. Placing from fifteen to twenty drops of water in one of these the stick of ink should be rubbed on the saucer with *moderate* pressure.

To properly mix ink requires great patience, as with too great pressure a mixture results having flakes and sand-like particles of ink in it, whereas an absolutely smooth and rather thick, slow-flowing liquid is wanted, whose surface will reflect the face like a mirror. The final test as to sufficiency of grinding is to draw a broad line and let it dry. It should then be a rich jet black, with a slight lustre. The end of the cake must be carefully dried on removing it from the saucer to prevent its flaking, which it will otherwise invariably do.

One may say, almost without qualification, and particularly when for use on tracing-cloth, the thicker the ink the better; but if it should require thinning, on saving it from one day to another — which is possible with the close-fitting saucers described — add a few drops of water, or of ox-gall if for use on a glazed surface.

When the ink has once dried on the saucer no attempt should be made to work it up again into solution. Clean the saucer and start anew.

WATER COLORS.

56. The ordinary colored writing inks should never be used by the draughtsman. They lack the requisite "body" and are corrosive to the pen. Very good colored drawing inks are now manufactured for line work, but Winsor and Newton's water colors, in the form called "moist," and in "half-pans" are the best if not the most convenient, for color work either with pen or brush. Those most frequently employed in engineering and architectural drawing are Prussian Blue, Carmine, Light Red, Burnt Sienna, Burnt Umber, Vermilion, Gamboge, Yellow Ochre, Chrome Yellow, Payne's Gray and Sepia. For some of their special uses see Art. 73.

Although hardly properly called a color Chinese White may be mentioned at this point as a requisite, and obtainable of the same form and make as the colors above.

DRAWING-PINS.

57. Drawing-pins or thumb-tacks, for fastening paper upon the board, are of various grades, the best, and at present the cheapest, being made from a single disc of metal one-half inch in diameter, from which a section is partially cut, then bent at right angles to the surface, forming the point of the pin.

IRREGULAR CURVES.

58. Irregular or French curves, also called *sweeps*, for drawing non-circular arcs, are of great variety, and the draughtsman can hardly have too many of them. They may be either of pear wood or hard rubber. A thoroughly equipped draughting office will have a large stock of these curves, which may be obtained in sets, and are known as railroad curves, ship curves, spirals, ellipses, hyperbolas, parabolas and combination curves. Some very serviceable *flexible* curves are also in the market.

Fig. 30.

If but two are obtained (which would be a minimum stock for a beginner) the forms shown in Fig. 30 will probably prove as serviceable as any. When employing them for inked work the pen should be so turned, as it advances, that its blades will maintain the same relation (parallelism) to the edge of the guiding curve as they ordinarily do to the edge of the rule. And the student must content himself with drawing slightly less of the curve than might apparently be made with one setting of the sweep, such course being safer in order to avoid too close an approximation to angles in what should be a smooth curve. For the same reason, when placed in a new position, a portion of the irregular curve must coincide with a part of that last inked.

The pencilled curve is usually drawn free-hand, after a number of the points through which it should pass have been definitely located. In sketching a curve free-hand it is much more naturally and smoothly done if the hand is always kept on the concave side of the curve.

INDIA RUBBER.

59. For erasing pencil-lines and cleaning the paper india rubber is required, that known as "velvet" being recommended for the former purpose, and either "natural" or "sponge" rubber for the latter. Stale bread crumbs are equally good for cleaning the surface of the paper after the lines have been inked, but will damage pencilling to some extent.

One end of the velvet rubber may well be wedge-shaped in order to erase lines without damaging others near them.

INK ERASER.

60. The double-edged erasing knife gives the quickest and best results when an inked line is to be removed. The point should rarely be employed. The use of the knife will damage the paper more or less, to partially obviate which rub the surface with the thumb-nail or an ivory knife handle.

PROTRACTOR.

61. For laying out angles a graduated arc called a "protractor" is used. Various materials are employed in the manufacture of protractors, as metal, horn, celluloid, Bristol board and tracing paper. The two last are quite accurate enough for ordinary purposes, although where the utmost precision is required, one of German silver should be obtained, with a moveable arm and vernier attachment.

Fig. 31.

The graduation may advantageously be to half degrees for average work.

To lay out an angle (say 40°) with a protractor, the radius CH (Fig. 31) should be made to coincide with one side of the desired angle; the centre, C, with the desired vertex; and a dot made with the pencil opposite division numbered 40 on the graduated edge. The line MC, through this point and C, completes the construction.

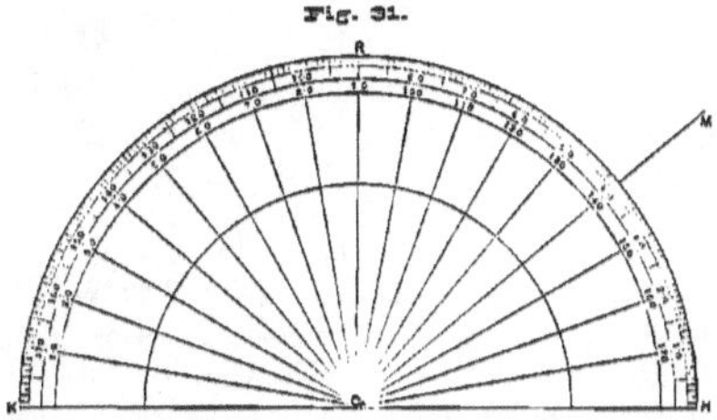

BRUSHES.

62. Sable-hair brushes are the best for laying flat or graduated tints, with ink or colors, upon small surfaces; while those of camel's hair, large, with a brush at each end of the handle, are better adapted for tinting large surfaces. Reject any brush that does not come to a perfect point on being moistened. Five or six brushes of different sizes are needed.

PRELIMINARIES TO PRACTICAL WORK.

63. The first work of a draughtsman, like most of his later productions, consists of *line* as distinguished from *brush* work, and for it the paper may be fastened upon the board with thumb-tacks only.

THEORETICAL AND PRACTICAL GRAPHICS.

There is no universal standard as to *size* of sheets for drawings. As a rule each draughting office has its own set of standard sizes, and system of preserving and indexing. The columns of the various engineering papers present frequent notes on these points, and the best system of preserving and recording drawings, tracings and corrections is apparently in process of evolution. For the student the best plan is to have all drawings of the same size bound in neat but permanent form at the end of the course. The title-pages, which presumably have also been drawn, will sufficiently distinguish the different sets.

In his elementary work the student may to advantage adopt two sizes of sheets which are considerably employed, $9'' \times 13''$, and its double, $13'' \times 18''$; sizes into which a "Super Royal" sheet naturally divides, leaving ample margins for the mucilage in case a "stretch" is to be made.

A "Double Elephant" sheet being twice the size of a "Super Royal" divides equally well into plates of the above size, but is preferable on account of its better quality.

To lay out four rectangles upon the paper locate first the centre (see Fig. 32) by intersecting diagonals, as at O. These should *not* be drawn entirely across the sheet, but one of them will necessarily pass a *short* distance each side of the point where the centre lies—judging by the eye alone; the second definitely determines the point. If the T-rule will not reach diagonally from corner to corner of the paper (and it usually will not) the edge may be practically extended by placing a triangle against but *projecting beyond* it, as in the upper left-hand portion of the figure.

Fig. 32.

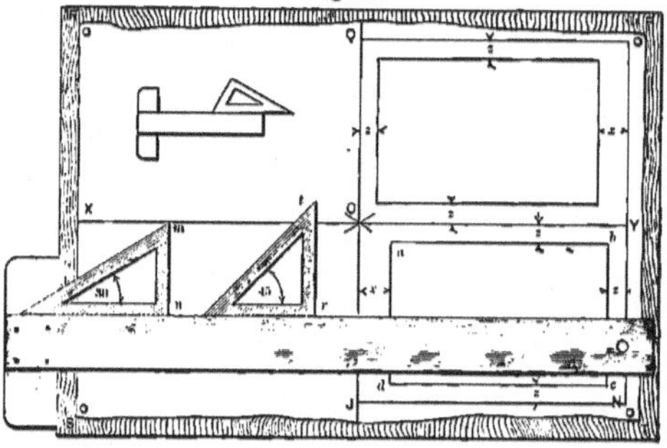

The T-rule being placed as shown, with its head *at the left end* of the board—the correct and usual position—draw a horizontal line XY, through the centre just located. The vertical centre line is then to be drawn, with one of the triangles placed as shown in the figure, i. e., so that a side, as mn or tr, is perpendicular to the edge.

It is true that as long as the edges of the board are exactly at right angles with each other we might use the T-rule altogether for drawing mutually perpendicular lines. This condition being, however, rarely realized for any length of time, it has become the custom—a safe one, as long as rule and triangle remain "true"—to use them as stated.

The outer rectangles for the drawings (or "plates," in the language of the technical school) are completed by drawing parallels, as JN and YN, to the centre lines, at distances from them of $9''$ and $13''$ respectively, laid off *from the centre*, O.

An inner rectangle, as $abcd$, should be laid out on each plate, with proper margins; usually at least an inch at the top, right and bottom, and an *extra half inch on the left* as an allowance for binding. These margins are indicated by x and z in the figure, as variables to which any convenient values may be assigned. The broad margin z in the upper rectangle will be at the draughtsman's left hand if he turns the board entirely around—as would be natural and convenient—when ready to draw on the rectangle QY.

CHAPTER IV.

GRADES OF LINES.—LINE TINTING.—LINE SHADING.—CONVENTIONAL SECTION-LINING.— FREQUENTLY RECURRING PLANE PROBLEMS.—MISCELLANEOUS PEN AND COMPASS EXERCISES.

65. Several kinds of lines employed in mechanical drawing are indicated in the figure below. While getting his elementary practice with the ruling-pen the student may group them as shown, or in any other symmetrical arrangement, either original with himself or suggested by other designs.

Fig. 33.

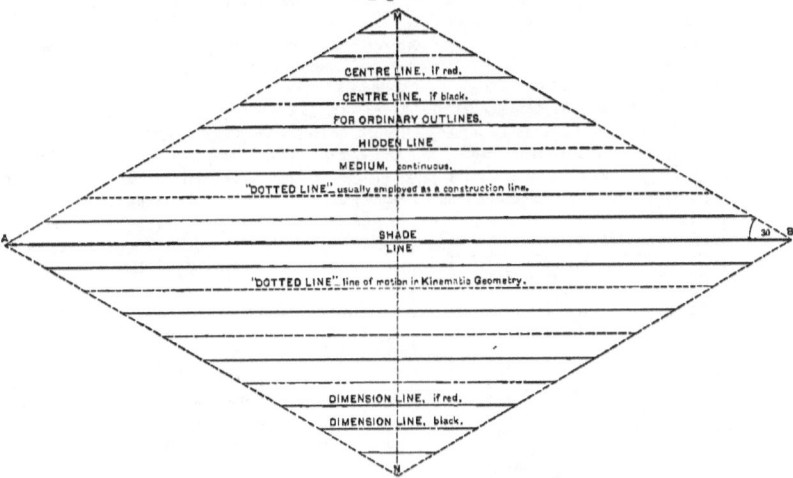

When drawing on tracing cloth or tracing paper, for the purpose of making blue-prints, all the lines will preferably be *black*, and the centre and dimension lines distinguished from others as indicated above, as also by being somewhat finer than those employed for the light outlines of the object. Heavy, opaque, *red* lines may, however, be used, as they will blue-print, though faintly.

There is at present no universal agreement among the members of the engineering profession as to standard dimension and centre lines. Not wishing to add another to the systems already at variance, but preferring to facilitate the securing of the uniformity so desirable, I have presented those for some time employed by the Pennsylvania Railroad and now taught at Cornell University.

The lines of Fig. 33, as also of nearly all the other figures of this work, having been printed from blocks made by the cerographic process (Art. 277), are for the most part too light to serve as examples for machine-shop work. Fig. 39 is a sample of P. R. R. drawing, and is a fair model as to weight of line for working drawings.

22   THEORETICAL AND PRACTICAL GRAPHICS.

A dash-and-three-dot line (not shown in the figure) is considerably used in Descriptive Geometry, either to represent an *auxiliary* plane or an *invisible* trace of *any* plane. (See Fig. 238).

The so-called "dotted" line is actually composed of short dashes. Its use as a "line of motion" was suggested at Cornell.

When colors are used without intent to blue-print they may be drawn as light, continuous lines. Colors will further add to the intelligibility of a drawing if employed for *construction* lines. Even if red dimension lines are used the *arrow heads* should invariably be *black*. They should be drawn *free-hand*, with a writing pen, and their points *touch* the lines between which they give the distance.

66. The utmost accuracy is requisite in pencilling, as the draughtsman should be merely a copyist when using the pen. On a complicated drawing even the *kind* of line should be indicated at the outset, so that no time will be wasted, when inking, in the making of distinctions to which thought has already been given during the process of construction. No unnecessary lines should be drawn, or any exceeding of the intended limit of a line if it can possibly be avoided.

If the work is symmetrical, in whole or in part, draw centre lines first, then main outlines; and continue the work from large parts to small.

The visible lines of an object are to be drawn first; afterward those to be indicated as concealed.

All lines of the same quality may to advantage be drawn with one setting of the pen, to ensure uniformity; and the light outlines before the shade lines.

In drawing arcs and their tangents ink the former first, invariably.

All the inking may best be done at once, although for the sake of clearness, in making a large and complicated drawing, a portion—usually the nearest and visible parts—may be inked, the drawing cleaned, and the pencilling of the construction lines of the remainder continued from that point.

The inking of the centre, dimension and construction lines naturally follows the completion of the main design.

67. In Fig. 34 we have a straight-line design usually called the "Greek Fret," and giving the student his first illustration of the use of the "shade line" to bring a drawing out "in relief." The law of the construction will be evident on examination of the numbered squares.

Without entering into the theory of shadows at this point we may state briefly the "shop rule" for drawing shade lines, viz., *right-hand and lower*. That is, of any pair of lines making the same turns together or representing the limit of the same flat surface, the *right-hand* line is the heavier if the pair is vertical, but the *lower* if they run horizontally; always subject, however, to the proviso that the line of intersection of two illuminated planes is never a shade line.

68. The conic section called the *parabola* furnishes another interesting exercise in ruled lines, when it is represented by its tangents as in Fig. 35. The angle CAE may be assumed at pleasure, and on the finished drawing the numbers may

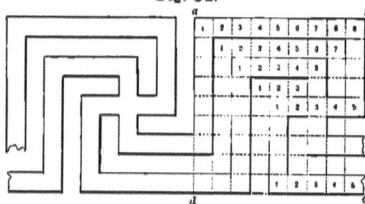

be omitted, being given here merely to show the law of construction. All the divisions are equal, and like numbers are joined.

Some interesting mathematical properties of the curve will be found in Chapter V.

69. A pleasing design that will test the beginner's skill is that of Fig. 36. It is suggestive of a cobweb, and a skillful free-hand draughtsman could make it more realistic by adding the spider. Use the 60° triangle for the heavy diagonals and parallels to them; the T-rule for the horizontals. Pencil the diagonals first but ink them last.

Fig. 36.

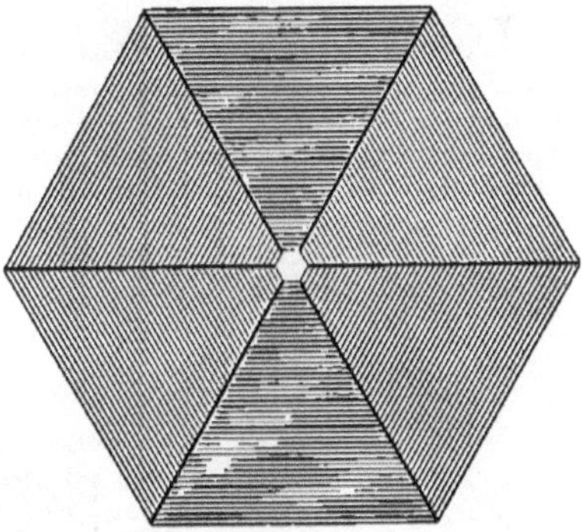

70. The even or flat effect of equidistant parallel lines is called *line-tinting;* or, if representing an object that has been cut by a plane, as in Fig. 37, it is called *section-lining.*

The *section*, strictly speaking, is the part actually in contact with the cutting plane; while the drawing as a whole is a *sectional view*, as it also shows what is back of the plane of section, the latter being always assumed to be transparent.

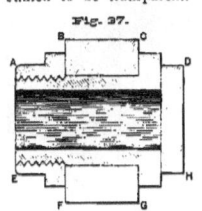

Fig. 37.

Adjacent pieces have the lines drawn in different directions in order to distinguish sufficiently between them.

The curved effect on the semi-cylinder is evidently obtained by properly varying both the strength of the line and the spacing.

71. The difference between the shading on the exterior and interior of a cylinder is sharply contrasted in Fig. 38. On the concavity the darkest line is at the top, while on the convex surface it is near the bottom, and below it the *spaces* remain unchanged while the lines diminish. A better effect would have been obtained in the figure had the engraver begun to increase the lines with the first decrease in the space between them.

Fig. 38.

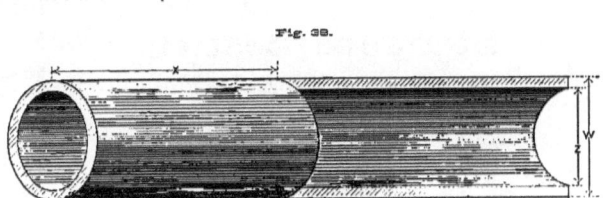

The spacing of the lines, in section-lining, depends upon the scale of the drawing. It may run down to a thirtieth of an inch or as high as one-eighth; but from a twentieth to a twelfth of an inch would be best adapted to the ordinary range of work. *Equal* spacing and not fine spacing

24 THEORETICAL AND PRACTICAL GRAPHICS.

should be the object, and neither scale nor patent section-liner should be employed, but distances gauged by the eye alone.

72. A refinement in execution which adds considerably to the effect is to leave a white line between the top and left-hand outlines of each piece and the section lines. When purposing to produce this effect rule light pencil lines as limits for the line-tints.

73. If the various pieces shown in a section are of different materials there are four ways of denoting the difference between them:

(a) By the use of the brush and certain water-colors, a method considerably employed in Europe, but not used to any great extent in this country, probably owing to the fact that it is not applicable where blue-prints of the original are desired.

The use of colors may, however, be advantageously adopted when making a highly finished, shaded drawing; the shading being done first, in India ink or sepia, and then overlaid with a flat tint of the conventional color. The colors ordinarily used for the metals are

        Payne's gray or India ink   for Cast Iron.
        Gamboge   " Brass (outside view).
        Carmine   " Brass (in section).
        Prussian Blue   " Wrought Iron.
        Prussian Blue with a tinge of Carmine " Steel.

Cast Iron.

Steel.

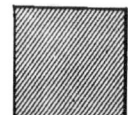
Wr't. Iron.

PENNA. R. R.

Standard Sections.

Brass.

Stone.

Wood.

Copper.

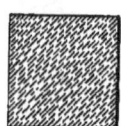
Brick.

CONVENTIONAL SECTION-LINING.

More natural effects can also be given by the use of colors, in representing the other materials of construction; and the more of an artist the draughtsman proves to be the closer can he approximate to nature.

Pale blue may be used for *water lines;* Burnt Sienna, whether grained or not, suggests *wood;* Burnt Umber is ordinarily employed for *earth;* either Light Red or Venetian Red are well adapted for *brick*, and a wash of India ink having a tinge of blue gives a fair suggestion of *masonry;* although the actual tint and surface of any rock can be exactly represented after a little practice with the brush and colors. These points will be enlarged upon later.

(b) By section-lining with the drawing pen in the conventional colors just mentioned, a process giving very handsome and thoroughly intelligible results on the original drawing, but, as before, unadapted to blue-printing and therefore not as often used as either of the following methods.

(c) By section-lining uniformly in *ink* throughout and printing the name of the material upon each piece.

(d) By alternating light and heavy, continuous and broken lines according to some law. Said "law" is, unfortunately, by no means universal, despite the attempt made at a recent convention of the American Society of Mechanical Engineers to secure uniformity. Each draughting office seems at present to be a law unto itself in this matter.

74. As affording valuable examples for further exercise with the ruling pen the system of section-lines adopted by the Pennsylvania Railroad is presented on the opposite page. The wood section is an exception to the rule, being drawn free-hand, with a Falcon pen.

By way of contrasting free-hand with mechanical work Fig. 40 is introduced, in which the rings showing annual growth are drawn as concentric circles with the compass.

In Fig. 41 a few other sections appear, selected from the designs of M. N. Forney and F. Van Vleck, and which are fortunate arrangements.

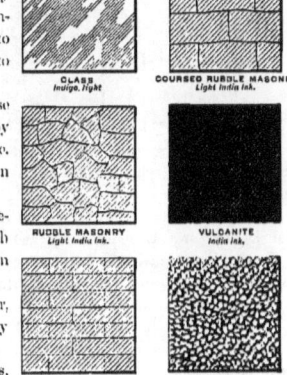

Fig. 41.

75. Figs. 42 and 43 are profiles or outlines of mouldings, such as are of frequent occurrence in architectural work. It is good practice to convert such views into oblique projections, giving the effect of solidity; and to further bring out their form by line shading. Figs. 44–46 are such representations, the front of each being of the same *form* as Fig. 42. The oblique lines are all parallel to each other, and—where

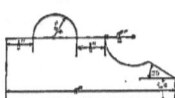
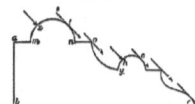
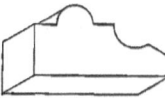

visible throughout—of the same length. Their *direction* should be chosen with reference to best exhibiting the peculiar features of the object. Obviously the view in Fig. 44 is the least adapted to the conveying of a clear idea of the moulding, while that of Fig. 46 is evidently the best.

76. The student may, to advantage, design profiles for mouldings and line-shade them, after converting them into oblique views. As hints for such work two figures are given (47-48), taken

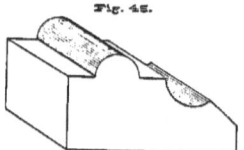
Fig. 45.

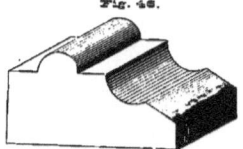
Fig. 46.

from actual construction in wood. By setting a moulding vertically, as in Fig. 49, and projecting horizontally from its points, a front view is obtained, as in Fig. 50.

Fig. 49.

Fig. 47.   Fig. 48.   Fig. 50.

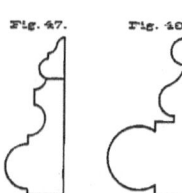
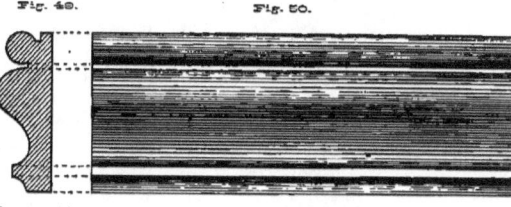

77. The *reverse curves* on the mouldings may be drawn with the irregular curve, (see Art 58); or, if composed of circular arcs to be tangent to vertical lines, by the following construction:—

Let M and N be the points of tangency on the verticals Mm and Nn, and let the arcs be tangent to each other at the middle point of the line MN. Draw Mn and Nm perpendicular to the vertical lines. The centres, c and c_1, of the desired arcs, are at the intersection of Mn and Nm by perpendiculars to MN from x and y, the middle points of the segments of MN.

Fig. 51.

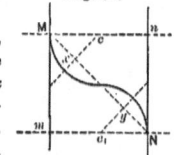

78. The light is to be assumed as coming in the usual direction, i. e., descending from left to right at such an angle that any ray would be projected on the paper at an angle of 45° to the horizontal.

In Fig. 48 several rays are shown. At x, where the light strikes the cylindrical portion most directly—technically is *normal* to the surface—is actually the brightest part. A tangent ray st gives t, the darkest part of the cylinder. The concave portion beginning at o would be darkest at o and get lighter as it approaches y.

Flat parts are either to be left white, if in the light, or have equidistant lines if in the shade, unless the most elegant finish is desired, in which case both change of space and gradation of line must be resorted to as in Fig. 52, which represents a front view of a hexagonal nut. The front face, being parallel to the paper, receives an even tint. An inclined face *in the light*, as $abhf$, is lightest toward the observer, while an unillumined face $tkdg$ is exactly the reverse.

Fig. 52.
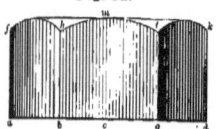

Notice that to give a *flat* effect on the inclined faces the spacing-out as also the change in the size of lines must be more gradual than when indicating curvature. (Compare with Figs. 46 and 50.)

REMARKS ON SHADING.—PLANE PROBLEMS.

If two or more illuminated flat surfaces are parallel to the paper (as *t g b h*, Fig. 52) but at different distances from the eye, the nearest is to be the lightest; if unilluminated, the reverse would be the case.

79. In treating of the theory of shadows distinctions have to be made, not necessary here, between *real* and *apparent* brilliant points and lines. We may also remark at this point that to an experienced draughtsman some license is always accorded, and that he can not be expected to adhere rigidly to theory when it involves a sacrifice of effect. For example, in Fig. 46 we are unable to see to the left of the (theoretically) lightest part of the cylinder, and find it, therefore, advisable to move the darkest part past the point where, according to Fig. 43, we know it in reality to be. The professional draughtsmen who draw for the best scientific papers and to illustrate the circulars of the leading machine designers allow themselves the latitude mentioned, with most pleasing results. Yet until one may be justly called an expert he can depart but little from the narrow confines of theory without being in danger of producing decidedly peculiar effects.

80. As from this point the student will make considerable use of the compasses, a few of the more important and frequently recurring plane problems, nearly all of which involve their use, may well be introduced. The proofs of the geometrical constructions are in several cases omitted, but if desired the student can readily obtain them by reference to any synthetic geometry or work on plane problems.

All the problems given (except No. 20) have proved of value in shop practice and architectural work.

The student should again read Arts. 48–51 regarding special uses of the 30° and 45° triangles, which, with the T-rule, enable him to employ so many "draughtsman's" as distinguished from "geometrician's" methods; also Arts. 36 and 37.

81. *Prob. 1.* To draw a perpendicular to a given line at a given point, as *A* (Fig. 53), use the triangles, or triangle and rule as previously described; or lay off equal distances *A a*, *A b*, and with *a* and *b* as centres draw arcs *o s t*, *m s n*, with common radius greater than one-half *a b*. The required perpendicular is the line joining *A* with the intersection of these arcs.

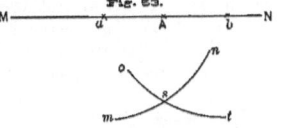

Fig. 53.

82. *Prob. 2.* To bisect a line, as *M N*, use its *extremities* exactly as *a* and *b* were employed in the preceding construction, getting also a second pair of arcs (same radius for all the arcs) intersecting *above* the line at a point we may call *x*. The line from *s* to *x* will be a bisecting perpendicular.

83. *Prob. 3.* To bisect an angle, as *A V B*, (Fig. 54), lay off on its sides any equal distances *V a*, *V b*. Use *a* and *b* as centres for intersecting arcs having a common radius. Join *V* with *x*, the intersection of these arcs, for the bisector required.

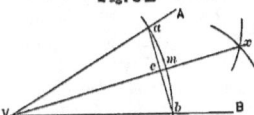

Fig. 54.

84. *Prob. 4.* To bisect an arc of a circle, as *a m b* (Fig. 54), bisect the chord *a c b* by Prob. 2; or, by Prob. 3, bisect the angle *a V b* which subtends the arc.

85. *Prob. 5.* To construct an angle equal to a given angle, as *θ* (Fig. 55), draw any arc *a b* with centre *O*, then, with same radius, an indefinite arc *m B*, centre *V*; use the *chord* of *a b* as a radius, and from centre *B* cut the arc *m B* at *x*. Join *V* and *x*. Then angle *A V B* equals *θ*.

Fig. 55.

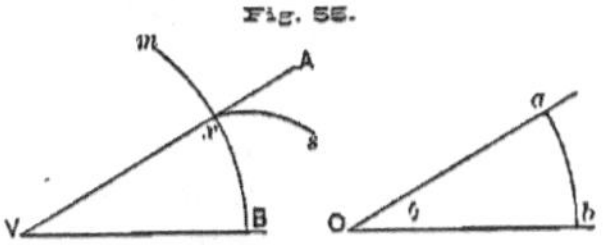

86. *Prob. 6.* To pass a circle through three points, *a*, *b* and *c*, join them by lines *a b*, *b c*, bisect these lines by perpendiculars, and the intersection of the latter will be the centre of the desired circle.

87. *Prob. 7.* To *divide a line into any number of equal parts* draw from one extremity as A

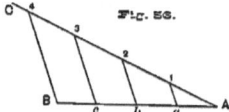
Fig. 56.

(Fig. 56) a line AC making any random angle with the given line AB. With a scale point off on AC as many equal parts (size immaterial) as are required on AB; four, for example. Join the last point of division (4) with B; then parallels to such line from the other points will divide AB similarly.

88. A *secant* to a curve is a line cutting it in two points. If the secant AB be turned to the left about A the point B will approach A, and the line will pass through AC and other secant positions. When B reaches and coincides with A the line is said to be *tangent* to the curve. (See also Art. 368.)

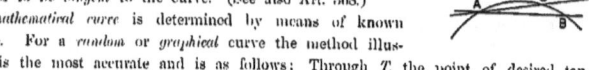
Fig. 57.

A *tangent* to a *mathematical curve* is determined by means of known properties of the curve. For a *random* or *graphical* curve the method illus-

Fig. 57 (a).

trated by Fig. 57 (a) is the most accurate and is as follows: Through T, the point of desired tangency, draw random secants to points on either side of it, as A, B, D, etc., and prolong them to meet a circle having centre T and *any* radius. On each secant lay off—from its intersection with the circle—the chord of that secant in the random curve. Thus, $am = TA$; $bn = TB$; $pd = TD$. From s, where the curve $mnopq$ cuts the circle, draw sT, which will be a tangent, since for it the chord has its minimum value.

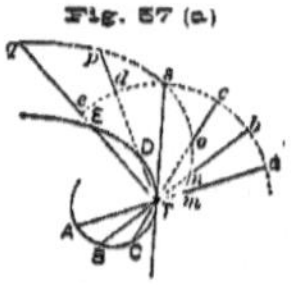

A *normal* to a curve is a line perpendicular to the tangent, at the point of tangency. In a circle it coincides in direction with the radius to the point of tangency.

89. *Prob. 8.* To *draw a tangent to a circle at a given point* draw a radius to the point. The perpendicular to this radius at its extremity will be the required tangent. Solve with triangles.

90. *Prob. 9.* To *draw a tangent to a circle from a point without* join the centre C (Fig. 58) with the given point A; describe a semicircle on AC as a diameter and join A with D, the intersection of the arcs. ADC equals 90°, being inscribed in a semi-circle; AD is then the required tangent, being perpendicular to CD at its extremity.

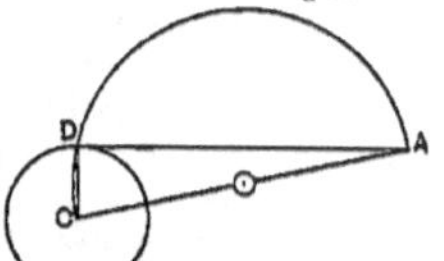

Fig. 58.

91. *Prob. 10.* To *draw a tangent at a given point of a circular arc whose centre is unknown or inaccessible*, locate *on the arc* two points equidistant from the given point and on opposite sides of it; the chord of these points will be parallel to the tangent sought.

92. A *regular* polygon has all its sides equal, as also its angles. If of three sides it is called the *equilateral triangle*; four sides, the *square*; five, *pentagon*; six, *hexagon*; seven, *heptagon*; eight, *octagon*; nine, *nonagon* or *enneagon*; ten, *decagon*; eleven, *undecagon*; twelve, *dodecagon*.

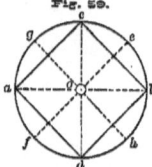
Fig. 59.

The angles of the more important regular polygons are as follows: *triangle*, 120°; *square*, 90°; *pentagon*, 72°; *hexagon*, 60°; *octagon*, 45°; *decagon*, 36°; *dodecagon*, 30°. The angle at the vertex of a regular polygon is the supplement of its central angle.

93. For the polygons most frequently occurring there are many special methods of construction. All but the pentagon and decagon can be readily inscribed or circumscribed about a circle by the use of the T-rule and triangle.

For example, draw ab (Fig. 59) with the T-rule, and cd perpendicular to it with a triangle. The 45° triangle will then give a *square*, $acbd$. The same triangle in two positions would give ef and gh, whence ag, gc, etc., would be sides of a regular *octagon*.

PLANE PROBLEMS.

94. The 60° triangle used as in Art. 51 would give the *hexagon;* and alternate vertices of the latter, joined, would give an *equilateral triangle.* Or the radius of the circle stepped off six times on the circumference, and alternate points connected, would result similarly.

Fig. 60.

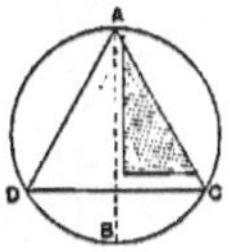

95. *Prob. 11.* An additional method for *inscribing an equilateral triangle in a circle, when one vertex of the triangle is given,* as A, Fig. 60. is to draw the diameter, AB, through A, and use the triangle to obtain the sides AC and AD, making angles of 30° with AB. D and C will then be the extremities of the third side of the triangle sought.

Fig. 61.

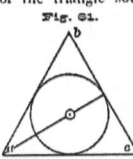

96. *Prob. 12. To inscribe a circle in an equilateral triangle* draw a perpendicular from any vertex to the opposite side. The centre of the circle will be on such line, two-thirds of the distance from vertex to base, while the radius desired will be the remaining third. (Fig. 61.)

97. *Prob. 13. To inscribe a circle in any triangle* bisect any two of the interior angles. The intersection of these bisectors will be the *centre*, and its perpendicular distance from any side will be the *radius* of the circle sought.

98. *Prob. 14. To inscribe a pentagon in a circle* draw mutually perpendicular diameters (Fig. 62); bisect a radius as at s; draw arc ax of radius sa and centre s; then chord $ax = af$, the side of the pentagon to be constructed.

Fig. 62.

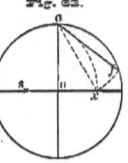

99. *Prob. 15. To construct a regular polygon of any number of sides, the length of the side being given.*

Let AB (Fig. 63) be the length assigned to a side, and a regular polygon of x sides desired. Take x equal to *nine* for illustration, draw a semi-circle with AB as radius and divide by trial into x (or 9) equal parts. Join B with $x-2$

Fig. 63.

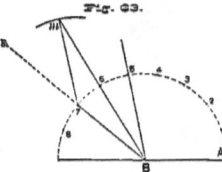

points of division, or *seven*, beginning at A, and prolong all but the last. With 7 as a centre, radius AB, cut line B-6 at m by an arc, and join m with 7, giving another side of the required polygon. Using m in turn as a centre, same radius as before, cut B-5 (produced) and so obtain a third vertex.

This solution is based on the familiar principles (a) that if a regular polygon has x sides each interior angle equals $\dfrac{180°(x-2)}{x}$, and (b) that the diagonals drawn from any vertex of the polygon make the same angles with each other as with the sides meeting at that vertex.

100. *Prob. 16. Another solution of Prob. 15.* Erect a perpendicular HR (Fig. 64) at the middle point of the given side. With M as a centre, radius MS, describe arc SA and divide it by trial into six equal parts. Arcs through these points of division, using A as a centre, and numbered up from six, give the centres on the vertical line for circles passing through M and S and in which MS would be a chord as many times as the number of the centre.

101. For any unusual number of sides the method

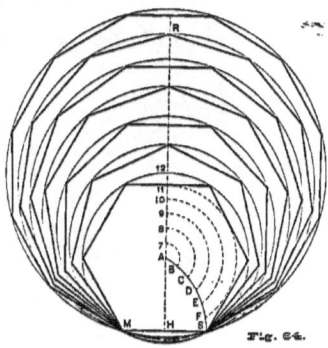

Fig. 64.

30 THEORETICAL AND PRACTICAL GRAPHICS.

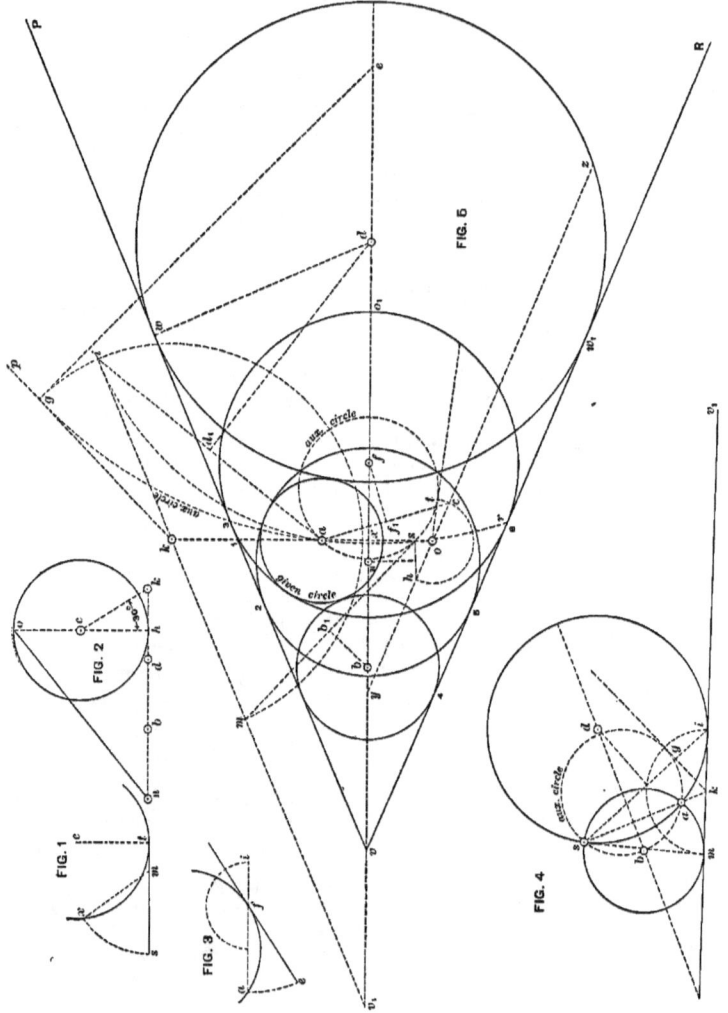

PLANE PROBLEMS.

by "trial and error" is often resorted to, and even for ordinary cases it is by no means to be despised. By it the dividers are set "by guess" to the probable chord of the desired arc, and, supposing a *heptagon* wanted, the chord is stepped off seven times around the circumference; care being taken to have the points of the dividers come exactly *on* the arc, and also to avoid damaging the paper. If the seventh step goes past the starting point the dividers require closing; if it falls short, the original estimate was evidently too small. Obviously the change in setting the dividers ought in this case to be, as nearly as possible, *one-seventh* of the error; and after a few trials one should "come out even" on the last step.

102. Prob. 17. *To lay off on a given circle an arc of the same length as a given straight line.*[1] Let t (Plate I, Fig. 1) be one extremity of the desired arc; ts the given straight line and tangent to the circle; tm equal one-fourth of ts, and sx drawn with centre m, radius ms. Then the length of the arc tx is a close approximation to that of the line ts.

103. Prob. 18. *To lay off on a straight line the length of a given circular arc,*[1] or, technically, to *rectify* the arc, let af (Plate I, Fig. 3) be the given arc; ai the chord prolonged till fi equals one-half the chord af; and ae an arc drawn with radius ai, centre i. Then fe approximates closely to the length of the arc af.

104. Prob. 19. *To obtain a straight line equal in length to any given semi-circle,*[2] draw a diameter oh of the given semi-circle (Plate I, Fig. 2) and a radius inclined at an angle of 30° to the radius ch. Prolong the radius to meet the line bhk, drawn tangent to the circle at h. From k lay off the radius three times, reaching n. The line no equals the semi-circumference to four places of decimals.

105. Prob. 20. *To draw a circle tangent to two straight lines and a given circle.* (Four solutions.) This problem is given more on account of the valuable exercise it will prove to the student in absolute precision of construction than for its probable practical applications. Fig. 4 (Plate I) illustrates the geometrical principles involved, and in it a circle is required to contain the points s and

---

[1] These methods of approximation were devised by Prof. Rankine. They are sufficiently accurate for arcs not exceeding 60°. The error varies as the fourth power of the angle. The complete demonstration of Prob. 17 can be found in the Philosophical Magazine for October, 1867, and of Prob. 18 in the November issue of the same year.

[2] In his *Graphical Statics* Cremona states this to be the simplest method known for rectifying a semi-circumference. According to Böttcher it is due to a Polish Jesuit, Kochansky, and was published in the Acta Eruditorum Lipsiae, 1685. The demonstration is as follows: Calling the radius *unity*, the diameter would have the numerical value 2.

Then in Fig. 2, Plate I, we have $on = \sqrt{oh^2 + hn^2} = \sqrt{oh^2 + (kn - kh)^2} = \sqrt{4 - (3 - \tan 30°)^2} = 3.14159 +$

The tangent of an angle (abbreviated to "tan.") is a trigonometric function whose numerical value can be obtained from a table. A draughtsman has such frequent occasion to use these functions that they are given here for reference, both as lines and as ratios.

Trigonometric Functions as Ratios.

θ = the given angle = CAB
h = hypothenuse of triangle CAB
$a = AB$ side of triangle *adjacent* to vertex of θ
$o = BC$ = side of triangle *opposite* to θ

Then $\sin \theta = \frac{o}{h}$; $\cos \theta = \frac{a}{h}$;

$\tan \theta = \frac{o}{a} = \frac{\sin \theta}{\cos \theta}$;

$\sec \theta = \frac{h}{a}$ = reciprocal of cosine.

$\csc \theta = \frac{h}{o}$ " " sine

$\cot \theta = \frac{a}{o} = \frac{\cos \theta}{\sin \theta}$ = reciprocal of tan θ.

Trigonometric Functions as Lines.

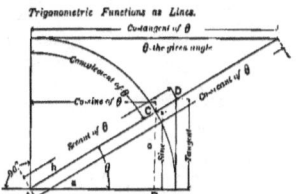

The prefix "co" suggests "complement;" the *co*-sine of θ is the sine of the complement of θ, &c. As *lines* the functions may be defined as follows:

The *sine* of an arc (o. g., that subtended by angle θ in the figure) is the perpendicular (CB) let fall from one extremity of the arc upon the diameter passing through the other extremity. If the radius AC, through one extremity of the arc, be prolonged to cut a line tangent at the other extremity, the intercepted portion of the tangent is called the *tangent* of the arc, and the distance, on such extended radius, from the centre of the circle to the tangent, is called the *secant of the arc*.

The co-sine, co-secant and co-tangent of the arc are respectively the sine, secant and tangent of the complement of the given arc.

a and be tangent to the line $m r_1$. Draw first any circle containing s and a, as the one called "aux. circle." Join s to a and prolong to meet $m v_1$ at k. From k draw a tangent, $k g$, to the auxiliary circle. With radius $k g$ obtain m and i on the line $m v$. A circle through s, a and m, or through s, a and i will fulfill the conditions. For $k g^2 = k s \times k a$, as $k g$ is a tangent and $k s$ a secant. But $k i = k g$, therefore $k i^2 = k s \times k a$, which makes $k i$ a tangent to a circle through s, a and i.

In Fig. 5 (Plate I) the construction is closely analogous to the above, and the lettering identical for the first half of the work. The "given circle" is so called in the figure; the given lines are $P r$ and $R r$. Having drawn the bisector, $v e$, of the angle $P v R$, locate s as much below $v e$ as a (the centre of the given circle) is above it, the line $a s$ being perpendicular to $v e$. Draw $v_1 m k i$ parallel to $r p$ and at a distance from it equal to the radius of the given circle. Then s, a, k and $m v_1$ of Fig. 5 are treated exactly as the analogous points of Fig. 4, and a circle obtained (centre d) containing a, s and i. The required circle will have the same centre d but radius $d w$, shorter than the first by the distance $w i$. Treat s, a, and m, (Fig. 5), similarly, getting the smallest of the four possible circles.

The remaining solutions are obtained by using the points a and s again, but in connection with a line $y z$ parallel to $v R$ and inside the angle, again at a perpendicular distance from one of the given lines equal to the radius of the given circle.*

This problem makes a handsome plate if the given and required lines are drawn in black; the lines giving the first two solutions in red; the remaining construction lines in blue.

106. *Prob. 21. To draw a tangent to two given circles* (a problem that may occur in connecting band-wheels by belts) join their centres, c and o, (Fig. 67) and at s lay off $s m$ and $s n$ each equal to the radius of the smaller circle. Describe a semi-circle $o h k e$ on $o c$ as a diameter. Carry m and n to k and h, about o as a centre. Angles $c k o$ and $c h o$ are each $90°$, being inscribed in a semi-circle; and $c k$ is parallel to $a b$, which last is one of the required tangents; while $c h$ is parallel to $t x$, a second tangent. Two more can be similarly found.

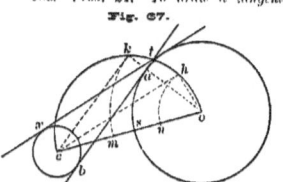

Fig. 67.

107. *Prob. 22. To unite two inclined straight lines by an arc tangent to both, radius given.* Prolong the given lines to meet at a (Fig. 68). With a as a centre and the given radius describe the arc $m n$. Parallels to the given lines and tangent to arc $m n$ meet at d, from which perpendiculars to the given lines give the points of tangency of the required arc, which is now drawn with the given radius.

Fig. 68.

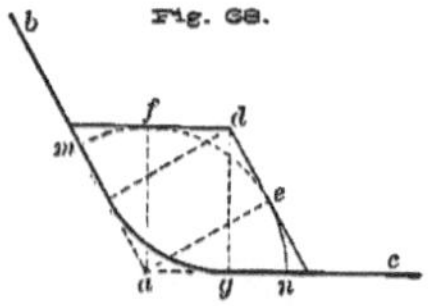

Fig. 69.

108. *Prob. 23. To draw through a given point a line which will—if produced—pass through the inaccessible*

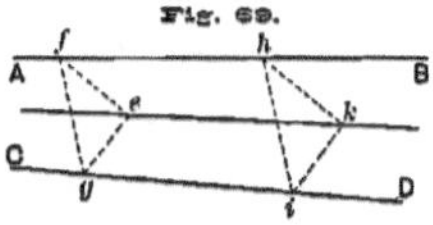

*This solution is taken from Benjamin Alvord's *Tangencies of Circles and of Spheres*, published by the Smithsonian Institute. That valuable pamphlet presents geometrical solutions of the ten problems of Apollonius on the tangencies of circles, and also of the fifteen problems on the tangencies of spheres, all of which are valuable to the draughtsman, both geometrically and as exercises in precision. The solutions are based on the principle, illustrated by Fig. 57, that the tangent line or tangent curve is the limit of all secant lines or curves. The problems on the tangencies of circles are as follows, the number of solutions in each case being given: (1) To draw a circle through three points. One solution. (2) Circle through two points and tangent to a given straight line. Two solutions. (3) Circle through a given point and tangent to two straight lines. Two solutions. (4) Circle through two points and tangent to a given circle. Two solutions. (5) Circle through a given point, tangent to a given straight line and a given circle. Four solutions. (6) Circle through a given point and tangent to two given circles. Four solutions. (7) Circle tangent to three straight lines, two only of which may be parallel. Four solutions. (8) Circle tangent to two straight lines and a given circle. Four solutions. (Art. 105, above). (9) Circle tangent to two given circles and a given straight line. Eight solutions. (10) Circle tangent to three given circles. Eight solutions.

intersection of two lines. Join the given point *c* with any point *f* on *A B* and also with some point *g* on *C D*. From any point *h* on *A B* draw *h i* parallel to *f g*, then *i k* parallel to *g c* and *h k* parallel to *f c*. The line *k e* will fulfill the conditions.

109. *Prob. 24. To draw an oval upon a given line.* Describe a circle on the given line, *m n*. (Fig

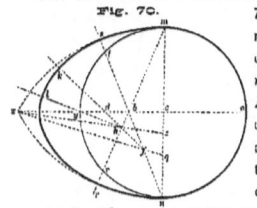

Fig. 70.

70) as a diameter. With *m* and *n* as centres describe arcs, *m x*, *n r* radius *m n*. Draw *m r* and *n t* through *r* and *t*, the middle points of the quadrants *y m*, *y n*. Then *m s* and *n r* are the portions of *m x* and *n x* forming part of the oval. Bisect *n c* at *q* and draw *q x*. Also bisect *c q* at *z* and join the latter with *x*. Bisect *y b* in *d* and draw *f d* from *f*, the intersection of *n s* and *q x*. Use *j* as a centre and *f s* as radius for an arc *s k* terminating on *f d*. The intersection, *h*, of *k f* with *x z* is then the next centre and *h k* the radius of the arc *k l* which terminates on *h y* produced. The oval is then completed with *y* as a centre and radius *y l*. The lower portion is symmetrical with the upper and therefore similarly constructed.

110. Where exact tangency is the requirement novices occasionally endeavor to conceal a failure to secure the desired object by thickening the curve. Such a course usually defeats itself and makes more evident the error they thus hope to conceal. With such instruments of precision as the draughtsman employs there is but little, if any, excuse for overlapping or falling short.

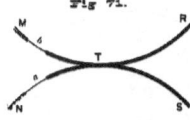

Fig. 71.

A common error in drawing tangents where the lines are of appreciable thickness is to make the outsides of the lines touch; whereas they should have their thickness in common at the point of tangency, as at *T* (Fig. 70), where, evidently, the centre-lines *a* and *b* of the arcs would be *exactly* tangent, while the outer arc of *M* would come tangent to the inner arc of *N*.

111. When either a tube or a solid cylindrical piece is seen in the direction of its axis the outline is, evidently, simply a circle; and often the only way to determine which of the two the circle represented would be to notice which part of said end view was represented as casting a shadow. In Fig. 72, if the shaded arcs can cast shadows, the space *inside* the circles must be open, and the figure would represent a portion of the end view of a boiler with its tubular openings.

Fig. 72.

By exactly reversing the shading, the effect of which can be seen by turning the figure upside down, it is converted into a drawing of a number of solid, cylindrical pins projecting from a plate.

The tapering begins at the points where a diameter at 45° to the horizontal would cut the circumference.

To get a perfect taper on small circles use the bow-pen and after making one complete circle add the extra thickness by a second turn which is to begin with the pen-point *in the air*, the pen being brought down gradually upon the paper and then, while turning, raised from it again.

On medium and large circles the requisite taper can be obtained by a different process, viz., by using the same radius again but by taking a *second centre*, distant from the first by an amount equal to the proposed width of the broadest part of the shaded arc; the line through the two centres to be perpendicular to that diameter which passes through the extremities of the taper.

112. As an exercise in concentric circles Fig. 73 will prove a good test of skill. It is a fair representation of a gymnasium ring, the "annular torus" of mathematical works, and possessing

Fig. 73.

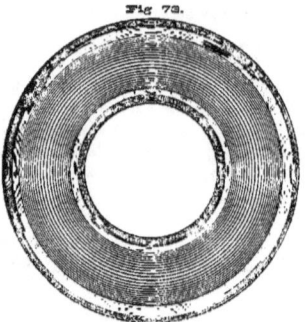

some remarkable properties, chief among which is the fact that it is the only surface of revolution known from which circles can be cut by three different systems of planes.*

Fig. 74. **Fig. 75.**

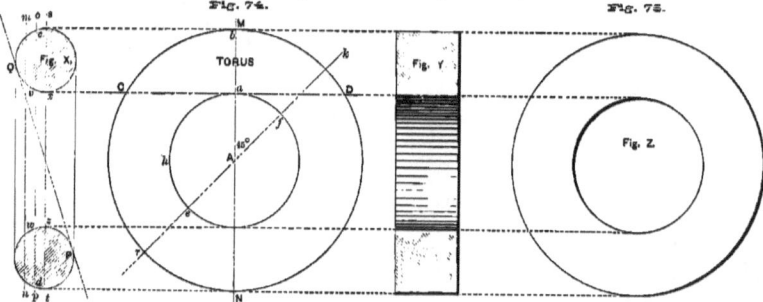

113. In Fig. 74 the same surface is shown, in the centre, in outline only. The axis of the surface would be a perpendicular to the paper at A. If MN represents a plane perpendicular to the paper and containing the axis, then Fig. X will show the shape of the cut or *section*. As MN was but one of the positions of a plane containing the axis and as the surface might be generated by rotating MN with the circle ab about the axis, it is evident that *one of the three systems of planes must contain the axis.*

When a surface can be generated by revolution about an axis one of its characteristics is that *any plane perpendicular to the axis will cut it in a circle.* The circles of Fig. 73 may then be, for the moment, considered as parallel cuts by a series of planes perpendicular to the axis, a few of which may be shown in mn, op, &c. (Fig. X). Each of these planes cuts two circles from the surface; the plane op, for example, giving circles of diameters cd and vw respectively.

* Olivier, *Mémoires de Géométrie Descriptive.* Paris, 1851.

ANNULAR TORUS.—WARPED HYPERBOLOID. 35

A plane perpendicular to the paper, on PQ, would be a *bi-tangent* plane, because tangent to the surface at two points. P and Q; and such plane would cut *two* over-lapping circles from the torus, each of them running partly on the inner and partly on the outer portion of the surface. For the proof that such sections are circles the student is probably not prepared at this point, but is referred to *Olivier's Seventh Memoir*.

114. Another interesting fact with regard to the torus is that a series of planes *parallel to, but not containing the axis*, cut it in a set of curves called the Cassian ovals, of which the Lemniscate of Art. 158 is a special case, and which would result from using a plane parallel to the axis and tangent to the surface at a point on the smallest circle, as at a, (Fig. 74.)*

115. Fig. Y is given to illustrate the fact that from mere untapered outlines, such as compose the central figure, we cannot determine the form of the object. By shading $e h f$ and $D N r$, the form shown in Fig. Y would be instantly recognized without the drawing of the latter. An angular object must therefore have shade lines, as also the *end view* of a round object; but a side view of a cylindrical piece must either have *uniform outlines* or be shaded with several lines.

Thus, in Fig. 76, A would represent an angular piece while B would indicate a circular cylinder; if *elliptical* its section would be drawn at one side as shown.

Fig. 76.

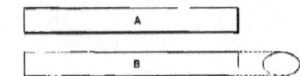

116. Before presenting the crucial test for the learner—the railroad rail—two additional practice exercises, mainly in ruling, are given in Figs. 77 and 78. The former shows that, like the parabola, the circle and hyperbola can be represented by their enveloping tangents. The upper and lower figures are merely two views of the surface called the *warped hyperboloid*, from the hyperbolas which constitute the curved outlines seen in the upper figure. The student can make this surface in a few moments by stringing threads through equidistant holes arranged in a circle on two circular discs, of the same or different sizes, but having the *same number of holes in each disc*. By attaching weights to the threads to keep them in tension at all times, and giving the upper disc a twist, the surface will change from cylindrical or conical to the hyperboloidal form shown.

Gear wheels are occasionally constructed, having their teeth upon such a surface and in the direction of the lines or elements forming it; but the hyperboloid is of more interest mathematically than mechanically.

Begin the drawing by pencilling the three concentric circles of the lower figure. When inking, omit the smaller circle. Draw a series of tangents to the inner circle, each one beginning on the middle circle and terminating on the outer. Assume any vertical height, $t s'$, for the upper figure, and draw $H' M'$ and $P' R'$ as its upper and lower limits. $H' M'$ is the vertical projection, or elevation, of the circle $H K M N$, and *all* points on the latter as $1, 2, 3, 4$, are projected, by perpendiculars to $H' M'$, at $1', 2', 3', 4'$, etc. All points on the larger circle $P Q R$ are similarly projected to $P' R'$. The extremities of the same tangent are then joined in the upper view, as $1'$ with 1 (a).

Part of each line is dotted to represent its disappearing upon an invisible portion of the surface. The law of such change on the lower figure is evident from inspection; while on the eleva-

---
* These curves can also be obtained by assuming two foci, as if for an ellipse, but taking the *product* of the focal radii as a constant quantity, some perfect square. If $pp' = 36''$ then a point on the curve would be found at the intersection of arcs having the foci as centres, and for radii $2''$ and $18''$, or $4''$ and $9''$, etc. When the constant assumed is the square of half the distance between the foci, the Lemniscate results.

36   THEORETICAL AND PRACTICAL GRAPHICS.

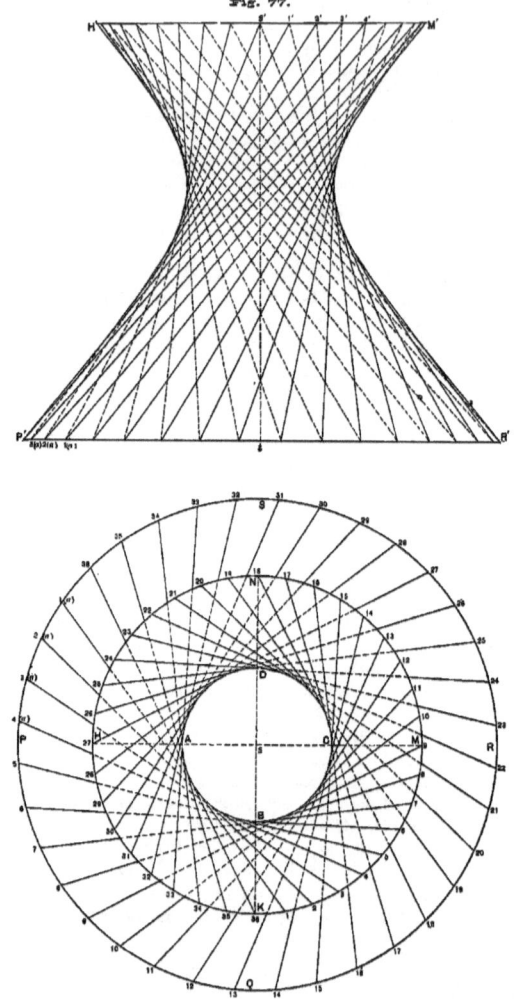

Fig. 77.

tion the point of division on each line is exactly above the point where the other view of the same line runs through HM in the lower figure.

117. To reproduce Fig. 78 draw first the circle $afbn$, then two circular arcs which would contain a and b if extended, and whose greatest distance from the original circle is x, (arbitrary). Sixteen equidistant radii as at a, c, d, etc., are next in order, of which the rule and 45° triangle give those through a, d, f and h. At their extremities, as m and n, lay off the desired width, y, and draw toward the points thus determined lines radiating from the centre. Terminate these last upon the inner arcs. Ink by drawing *from* the centre, *not through* or *toward* it.

All construction lines should be erased before the tapering lines are filled in. The "filling in" may be done very rapidly by ruling the edges of the line in *fine* at first, then opening the pen slightly and beginning again where the opening between the lines is apparent and ruling from there, adding thickness to each edge on its inner side. It will then be but a moment's work to fill in, free-hand, with the Falcon pen or a fine-pointed sable-brush, between the now heavy edge-lines of the taper. To have the pen make a coarse line when starting from the centre would destroy the effect desired.

Fig. 78.

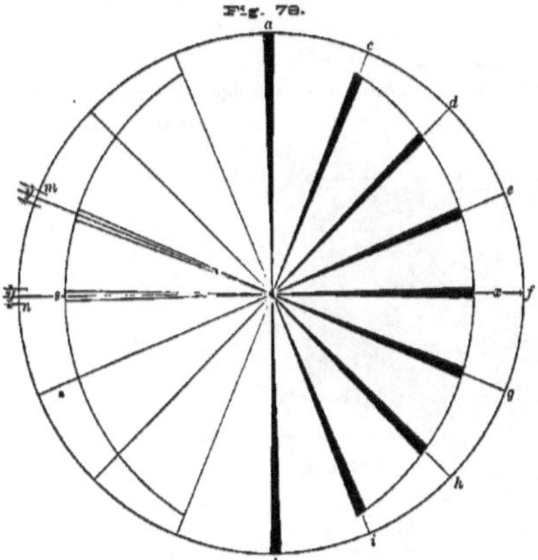

118. The draughtsman's ability can scarcely be put to a severer test on mere outline work than in the drawing of a railroad rail, so many are the changes of radii involved.

As previously stated, where tangencies to straight lines are required, the arcs are to be drawn first, then the tangents.

Figs. 79 and 80 are photo-engravings of rail sections, showing two kinds of "finish." Fig. 80 is a "working drawing" of a Pennsylvania Railroad rail, full-size. If finished with shade lines, as in Fig. 79, section-lined with Prussian blue, and the dimension lines drawn in carmine this makes one of the handsomest plates that can be undertaken.

A still higher effect is shown in the wood-cut of the title-page, the rail being represented in oblique projection and shaded.

Begin Fig. 80 by drawing the vertical centre-line, it being an axis of symmetry. Upon it lay off 5" for the total height, and locate two points between the top and base at distances from them of $1\frac{1}{4}$" and $\frac{3}{4}$" respectively; these to be the points of convergence of the lower lines of the head and sloping sides of the base. From these points draw lines, at first indefinite in length, and inclined 13° to the horizontal. The top of the head is an arc of 10" radius, subtended by an angle of 9°. This changes into an arc of $\frac{1}{16}$" radius on the upper corner, with its centre on the side of said 9° angle. The sides of the head are straight lines, drawn at 4° to the vertical and tangent to the corner arcs. The thin vertical portion of the rail is called the *web* and is $\frac{1}{2}$" wide at its centre. The outlines of the web are arcs of 8" radius, subtended by angles of 15°, centres on line marked "centre line of bolt holes."

The weight per yard of the rail shown is given as eighty-five pounds, from which we know the area of the cross-section to be eight and one-half square inches, since a bar of iron a yard long and one square inch in cross section weighs, approximately, ten pounds. (10.2 lbs., average.)

38 *THEORETICAL AND PRACTICAL GRAPHICS.*

The proportions given are slightly different from those recommended in the report* of the committee appointed by the American Society of Civil Engineers to examine into the proper relations to

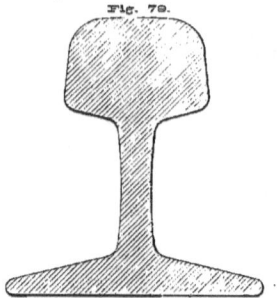

each other of the sections of railway wheels and rails. There was quite general agreement as to the following recommendations: a top radius of twelve inches; a quarter-inch corner radius; vertical sides

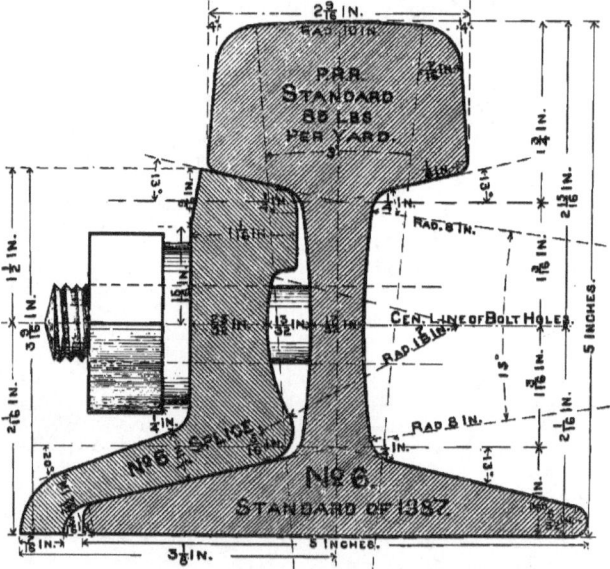

to the web; a lower-corner radius of one-sixteenth inch, and, lastly, a broad head relatively to the depth.

*Transactions A. S. C. E. January, 1891.

EXERCISES FOR THE IRREGULAR CURVE. 39

CHAPTER V.

THE HELIX.—CONIC SECTIONS.—HOMOLOGICAL PLANE CURVES AND SPACE-FIGURES.—LINK-MOTION CURVES.—CENTROIDS.—THE CYCLOID.—COMPANION TO THE CYCLOID.—THE CURTATE TROCHOID.—THE PROLATE TROCHOID.—HYPO-, EPI-, AND PERI-TROCHOIDS.—SPECIAL TROCHOIDS—ELLIPSE, STRAIGHT LINE, LIMAÇON, CARDIOID, TRISECTRIX, INVOLUTE, SPIRAL OF ARCHIMEDES.—PARALLEL CURVES.—CONCHOID.—QUADRATRIX.—CISSOID.—TRACTRIX.—WITCH OF AGNISI.—CARTESIAN OVALS.—CASSIAN OVALS.—CATENARY.—LOGARITHMIC SPIRAL.—HYPERBOLIC SPIRAL.—THE LITUUS.

119. There are many curves which the draughtsman has frequent occasion to make whose construction involves the use of the irregular curve. The more important of these are the Helix; Conic Sections—Ellipse, Parabola and Hyperbola; Link-motion curves or point-paths; Centroids; Trochoids; the Involute and the Spiral of Archimedes. Of less practical importance, though equally interesting geometrically, are the other curves mentioned in the heading.

The student should become thoroughly acquainted with the more important geometrical properties of these curves, both to facilitate their construction under the varying conditions that may arise and also as a matter of education. Considerable space is therefore allotted to them here.

At this point Art. 58 should be reviewed, and in addition to its suggestions the student is further advised to work, at first, on as large a scale as possible, not undertaking small curves of sharp curvature until after acquiring some facility in the use of the curved ruler.

THE HELIX.

120. The ordinary helix is a curve which cuts all the elements of a cylinder at the same angle. Or we may define it as the curve which would be generated by a point having a uniform motion around a straight line combined with a uniform motion parallel to the line.

Fig. 81.

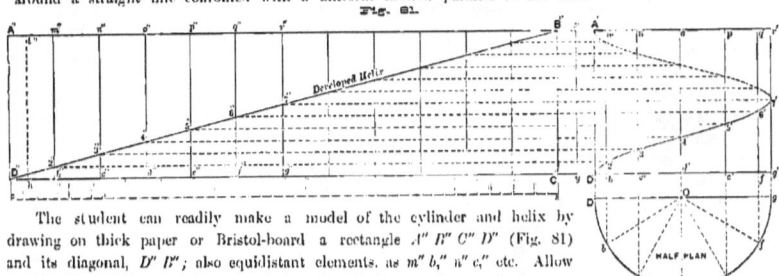

The student can readily make a model of the cylinder and helix by drawing on thick paper or Bristol-board a rectangle $A'' B'' C'' D''$ (Fig. 81) and its diagonal, $D'' B''$; also equidistant elements, as $m'' b''$, $n'' c''$, etc. Allow at the right and bottom about a quarter of an inch extra for overlapping, as shown by the lines $x y$ and $s z$. Cut out the rectangle $z x$; also cut a series of vertical slits between $D'' C''$ and $z s$ and turn the divisions up at right angles to the paper; put mucilage between $B'' C''$ and $x y$; then roll the paper up into cylindrical form, bringing $A'' D'' C'' h''$ in front of and upon the gummed portion, so that $A'' D''$ will coincide with $B'' C'''$. The diagonal $D'' B''$ will then be a helix on the outside of the cylinder, but half of which can be seen in front view, as D' $7'$, (see right-hand figure); the other half, $7' A'$, being indicated as unseen.

To give the cylinder more permanent form it can be pasted to a cardboard base by mucilage on the under side of the divisions turned up along its lower edge.

The rectangle $A'' B'' C'' D''$ is called the *development* of the cylinder; and any surface like a cylinder or cone, which can be rolled out on a plane surface and its equivalent area obtained by bringing consecutive elements into the same plane, is called a *developable surface*. The elements $m'' b''$, $n'' c''$, etc., of the development stand vertically at $b, c, d \ldots g$ of the half plan, and are seen in the elevation at $m' b', n' c', o' d'$, etc. The point $3'$, where any element, as c', cuts the helix, is evidently as high as $7'$ where the same point appears on the development. We may therefore get the curve $D' 7' A'$ by erecting verticals from $b, c, d, \ldots g$; to meet horizontals from the points where the diagonal D'' B'' crosses those elements on the development.

The length $D'' C'''$ obviously equals $2 \pi r$, in which $r = O D$.

The height $A' B'$, that the curve attains in winding once around the cylinder, is called the *pitch* of the helix.

The construction of the helix is involved in the designing of screws and screw-propellers, and in the building of winding stairs and skew-arches.

Mathematically the helix and its orthographic projection are well worth' study, particularly the latter when the helix crosses the elements at 45°; it becoming then identical with Roberval's *curve of sines*, otherwise known as the *companion to the cycloid*.

For a helix on a cone refer to the article on the Spiral of Archimedes.

THE CONIC SECTIONS.

121. The ellipse, parabola and hyperbola are called *conic sections* or *conics* because they may be obtained by cutting a cone by a plane. We will, however, first obtain them by other methods.

According to the definition given by Boscovich, the ellipse, parabola and hyperbola are curves in which there is a *constant ratio* between the distances of points on the curve from a certain fixed point (the *focus*) and their distances from a fixed straight line (the *directrix*).

Referring to the parabola, Fig. 82, if S and B are points of the curve, F the focus and $X Y$ the directrix, then, if $S F : S T :: B F : B X$, we conclude that B and S are points of a conic section.

122. The actual value of the ratio (or *eccentricity*) may be 1 or either greater or less than unity. When $S F$ equals $S T$ the ratio equals 1, and the relation is that of equality, or parity, which suggests the parabola.

123. If it is farther from a point of the conic to the focus than to the directrix the ratio is greater than 1 and the *hyperbola* is indicated.

124. The ellipse, of course, comes in for the third possibility as to ratio, viz., less than 1. Its construction by this principle is not shown in Fig. 82 but later, the method of generation here given illustrating the practical way in which, in landscape gardening, an elliptical plat would be laid out; it is therefore called the construction as the "gardener's ellipse."

Taking $A C$ and $D E$ as representing the extreme length and width, the points F and F_1 (*foci*) would be found by cutting $A C$ by an arc of radius equal to one-half $A C$, centre D. Pegs or pins at F and F_1, and a string, of length $A B$, with ends fastened at the foci, complete the preliminaries. The curve is then traced on the ground by sliding a pointed stake against the string, as at P, so that at all times the parts $F_1 P$, $F P$, are kept straight.

CONIC SECTIONS.

125. According to the foregoing construction the ellipse may be defined as a curve in which the sum of the distances from any point of the curve to two fixed points is constant. That constant is evidently the longer or *transverse* (*major*) axis, AC. The shorter or *conjugate* (or *minor*) axis, DE, is perpendicular to the other.

With the compasses we can determine P and other points of the ellipse by using F and F_1 as centres, and for radii any two segments of AC. Q, for example, gives AQ and CQ as segments. Then arcs from F and F_1, with radius equal to QC, would intersect arcs from the same centres, radius QA, in four points of the ellipse, one of which is P.

126. By the Boscovich definition, we are enabled to construct the parabola and hyperbola also by continuous motion along a string.

For the parabola place a triangle as in Fig. 82, with its altitude GX toward the focus. If a string *of length* GX be fastened at G, stretched tight from G to *any* point B, by putting a pencil at B, then the remainder BX swung around and the end fastened at F, it is then, evidently, as far from B to F as it is from B to the directrix; and that relation will remain constant as the triangle is slid along the directrix, if the pencil point remains against the edge of the triangle so that the portion of the string from G to the pencil is kept straight.

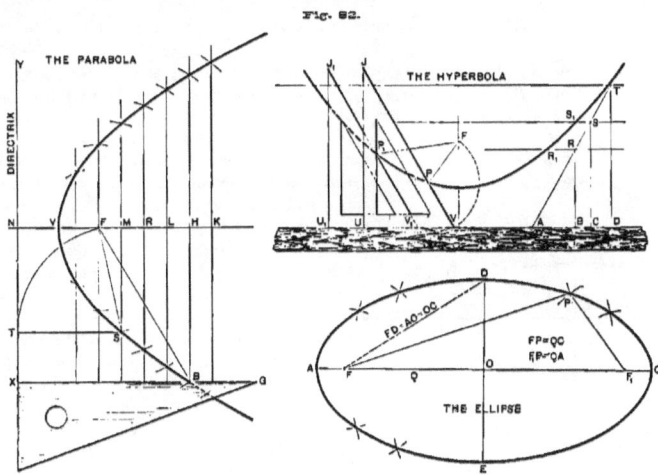

Fig. 82.

127. For the hyperbola, (Fig. 82), the construction is identical with the preceding, except that the string fastened at J runs down the *hypothenuse* and equals it in length.

128. Referring back to Fig. 35, it will be noticed that the focus and directrix of the parabola are there omitted; but the former would be the point of intersection of a perpendicular from A upon the line joining C with E. A line through A, parallel to CE, would be the directrix.

129. Like the ellipse, the hyperbola can be constructed by using two foci, but whereas in the ellipse (Fig. 82) it was the *sum* of two focal radii that was constant, i. e., $FP + F_1P = FD +$

$F_1 D = A C$ (the transverse axis), it is the *difference* of the radii that is constant for the hyperbola.

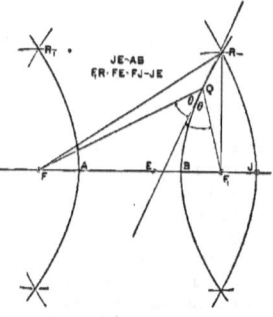

Fig. 83.

In Fig. 83, if $A B$ is to be the transverse axis of the two arcs, or "branches," which make the complete hyperbola, then using ρ and ρ' to represent any two focal radii, as $F Q$ and $F_1 Q$, or $F R$ and $F_1 R$, we will have $\rho - \rho' = A B$ (the constant quantity).

To get a point of the curve in accordance with this principle we may lay off from either focus, as F, any distance greater than $F B$, as $F J$, and with it as a radius, and F as a centre, describe the indefinite arc $J R$. Subtracting the constant, $A B$, from $F J$, by making $J E = A B$, we use the remainder, $F E$, as a radius, and F_1 as a centre to cut the first arc at R. The same radii will evidently determine three other points fulfilling the conditions.

130. The tangent to the hyperbola at any point, as Q, bisects the angle $F Q F_1$, between the focal radii.

In the ellipse, (Fig. 82), it is the *external* angle between the radii that is bisected by the tangent.

In the parabola, (Fig. 82), the same principle applies, but as one focus is supposed to be at *infinity*, the focal radius, $B G$, toward the latter, from any point, as B, would be parallel to the axis. The tangent at B would therefore bisect the angle $F B X$.

131. *The ellipse as a circle viewed obliquely.* If $A R M B F$ (Fig. 84) were a circular disc and we

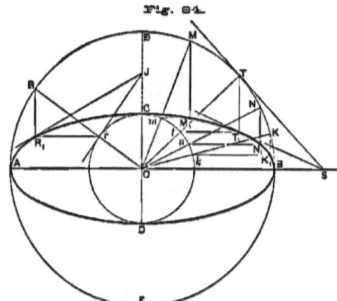

Fig. 84.

were to rotate it on the diameter $A B$, it would become narrower in the direction $F E$ until, if sufficiently turned, only an edge view of the disc would be obtained. The axis of rotation, $A B$, would, however, still appear of its original length. In the rotation supposed, all points not on the axis would describe circles about it with their planes perpendicular to it. M, for example, would move in the arc of a circle part of which is shown in $M M_1$, which is straight, as the plane of the arc is seen "edge-wise." If instead of a circular disc we were to take one of *elliptical* form, as $A C B D$, and turn it upon its shorter axis $C D$, it is obvious that B would apparently approach O on one side while A advanced on the other; and when B was directly in front of, and projected at k, we would have the ellipse projected in the small circle $r t k$. Having given then the two axes of a desired ellipse, as $A B$ and $C D$, use them as diameters of concentric circles and from their common centre, O, draw random radii, as $O M$, $O T$, $O K$. Where any radius, $O M$, meets the outer circle, drop a perpendicular $M M_1$, to the longer axis, to meet a line $m M_1$, parallel to the same longer axis and passing through m, where $O M$ cuts the smaller circle.

132. If $T S$ is a tangent at T to the large circle, then when T appears at T_1 we shall have $T_1 S$ as a tangent to the ellipse at the point derived from T; S having remained constant, being on the axis of rotation.

CONIC SECTIONS.

Similarly, if a tangent at R_1 were wanted we may first find r, corresponding to R_1; draw the tangent $r J$ to the small circle; then join J, (constant point), with R_1.

133. Occasionally we have given the length and inclination of a pair of diameters of the ellipse,

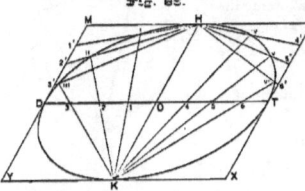

Fig. 85.

making oblique angles with each other. Such diameters are called *conjugate* and the curve may be constructed upon them thus: Draw the axes $T D$ and $H K$ at the assigned angle $D O H$; construct the parallelogram $M N$ $X Y$; divide $D M$ and $D O$ into the same number of equal parts; then from K draw lines through the points of division on $D O$, to meet similar lines through H and the divisions on $D M$. The intersection of like numbered lines will give points of the ellipse.

134. It is the law of expansion of a perfect gas that the volume is inversely as the pressure. That is, if the volume be doubled the pressure drops one-half; if trebled the pressure becomes one-third, etc. Steam not being a perfect gas departs somewhat from the above law, but the curve indicating the fall in pressure due to its expansion is compared with that for a perfect gas.

To construct the curve for the latter let us suppose $C L K G$ (Fig. 86) to be a cylinder with a volume of gas $C G b e$ behind the piston. Let $e b$ indicate the pressure before expansion begins. If the piston be forced ahead by the expanding gas until the volume is doubled, the pressure will drop, by Boyle's law, to one-half and will be indicated by $t d$. For three volumes the pressure becomes $v f$, etc. The curve $e s x$ is an hyperbola, the special case called *equilateral*.

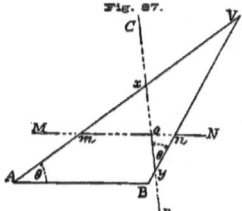

Fig. 86.

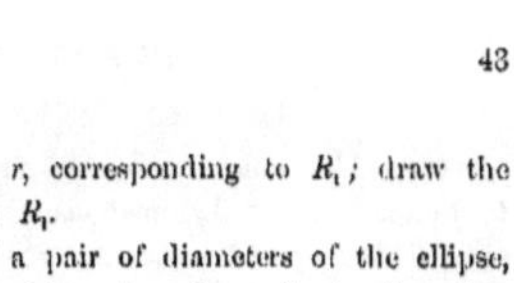

Suppose the cylinder were infinite in length. Since we cannot conceive a volume so great that it could not be doubled, or a pressure so small that it could not be halved, it is evident that theoretically the curve $e x$ and the line $G K$ will forever approach each other yet never meet; that is, they will be *tangent at infinity*. In such a case the straight line is called an *asymptote* to the curve.

135. Although the *right* cone (i. e., one having its axis perpendicular to the plane of its base) is usually employed in obtaining the ellipse, hyperbola and parabola, yet the same kind of

Fig. 87.

sections can be cut from an oblique or *scalene* cone of circular base, as $V A B$, Fig. 87. Two sets of *circular* sections can also be cut from such a cone, one set, obviously, by planes parallel to the base, the other by planes like $C D$, making the same angle with the lowest element, $V B$, that the highest element, $V A$, makes with the base. The latter sections are called *sub-contrary*.

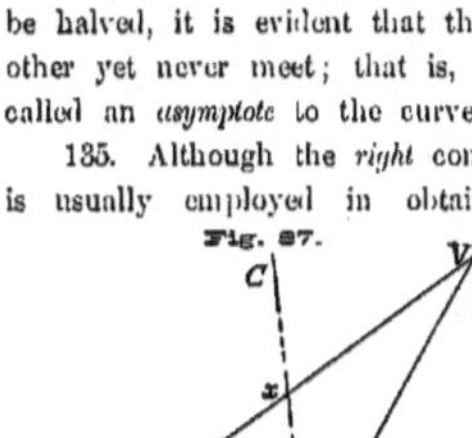

To prove that the section $x y$ is a circle we note that both it and the section $m n$—the latter known to be a circle because parallel to the base—intersect in a line perpendicular to the paper at o. This line pierces the front surface of the cone at a point we may call r. It would be seen as the ordinate $o r$ (Fig. 88), were the front half of circle

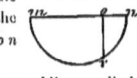

Fig. 88.

$m n$ rotated until parallel to the paper. Then $o r^2 = o m \times o n$. But from the similar triangles in Fig. 87 we have $o m : o y :: o x : o n$ whence $o y \times o z = o m \times o n = o r^2$, thus showing that the section $x y$ is circular.

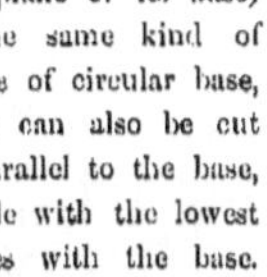

Were the vertex of a scalene cone removed *to infinity* the cone would become an oblique cylinder with circular base; but the latter would possess the property just established for the former.

44 THEORETICAL AND PRACTICAL GRAPHICS.

136. The most interesting practical application of the sub-contrary section is in Stereographic Projection, one of the methods of representing the earth's surface on a map. The especial convenience of this projection is due to the fact that in it every circle is projected as a circle. This results from the relative position of the eye (or centre of projection) and the plane of projection; the latter is that of some great circle of the earth and the eye is located at the pole of such circle.

137. In Fig. 89 let the circle ABE represent the equator; MN the plane of a meridian, also taken as the plane of projection; AB

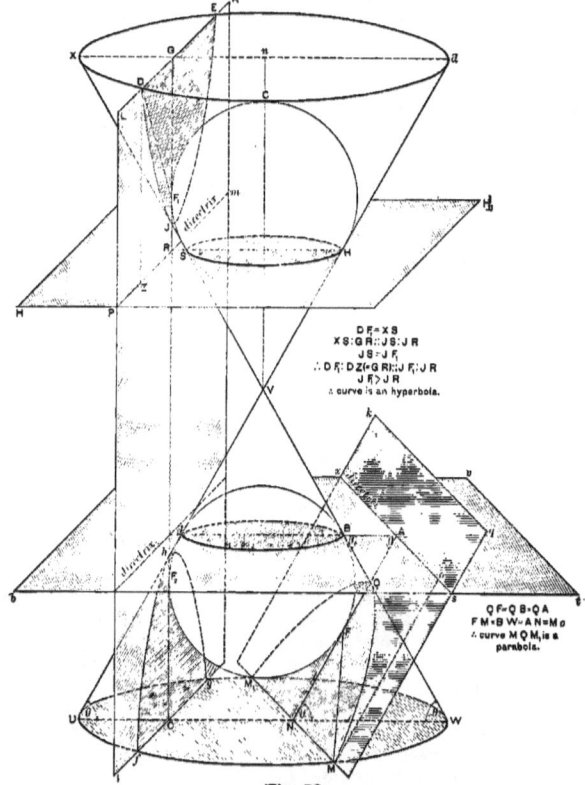

any circle of the sphere; E the position of the eye: then ab, the projection of AB on plane MN,

CONIC SECTIONS.

(see Art. 4), is a circle, being a sub-contrary section of the visual cone $E A B$, as the student can easily prove.

138. We now take up the conic sections as derived from a right cone.

A complete cone (Fig. 90) lies as much above as below the vertex. To use the term adopted from the French, it has two *nappes*.

Aside from the extreme cases of perpendicularity to or containing the axis *the inclination of a plane cutting the cone may be*

(a) *Equal to* that of the elements,* therefore parallel to *one* element, giving the *parabola*, as $M Q M_1$ (Fig. 90); the plane $k q M$ being parallel to the element $V U$ and therefore making with the base the same angle, θ, as the latter.

(b) *Greater than* that of the elements, causing the plane to cut both nappes and giving a two-branch curve, the *hyperbola*, as $D J E$ and $f h g$ (Fig. 90).

(c) *Less than* that of the elements, the plane therefore cutting all the elements on one side of the vertex, giving a closed curve, the *ellipse*; as $K s H$, Fig. 91.

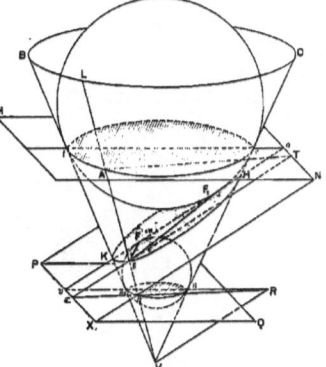

Fig. 91.

139. Figures 90 and 91, with No. 4 of Plate 2 are not only stimulating examples for the draughtsman but they illustrate probably the most interesting fact met with in the geometrical treatment of conic sections, viz., that the spheres which are tangent simultaneously to the cone and the cutting planes, touch the latter in the foci of the conics; while in each case the directrix of the curve is the line of intersection of the cutting plane and the plane of the circle of tangency of cone and sphere.

To establish this we need only employ the well known principles that (a) all tangents from a point to a sphere are equal in length, and (b) all tangents are equal that are drawn to a sphere from points equidistant from its centre. In both figures all points of the cone's bases are evidently equidistant from the centres of the tangent spheres.

140. On the upper nappe, (Fig. 90), let $S H$ be the circle of contact of a sphere which is tangent at F_1 to the cutting plane $P L K$. The plane $P H_0$ of the circle cuts the plane of section in $P m$. If D is *any* point of the curve $D J E$, J another point, and we can prove the ratio constant (and greater than unity) between the distances of D and J from F_1 and their distances to $P m$, then the curve $D J E$ must be an *hyperbola*, by the Boscovich definition; F_1 must be the focus and $P m$ the directrix.

$D F_1$ is a tangent whose real length is seen at $X S$. $J F_1$ and $J S$ are equal, being tangents to the sphere from the same point. We have then the proportion $X S : G R :: J S : J R$ or $D F_1 : D Z :: J F_1 : J R$. Since $J S$ and its equal $J F_1$ are greater than $J R$, and the ratio $J F_1$ to $J R$ is constant, the proposition is established.

141. For the parabola on the lower nappe, since the plane $M c k$ is inclined at the angle θ made by the elements, we have $Q A = Q B$ (opposite equal angles) and $Q B$ equal $Q F$ (equal tangents). $M F = B W = M o$, therefore $M F : M o :: Q F : Q o$, and it is as far from M to the focus F as to the directrix $s z$, fulfilling the condition essential for the parabola.

*See Remark, Art. 4.

46   *THEORETICAL AND PRACTICAL GRAPHICS.*

142. For the ellipse $K s H$, (Fig. 91), we have PX and NT as the lines to be proven directrices, and F and F_1 the points of tangency of two spheres. Let s be any point of the curve under consideration and VL the element containing s. This element cuts the contact circles of the spheres in a and A. A plane through the cone's axis and parallel to the paper would contain $o t$, $o v$ and $v n$. Prolong $v n$ to meet a line VR that is parallel to KH. Join R with a, producing it to meet PX at r. In the triangles $a s r$ and $a VR$ we have $s a : s r :: Va : VR$. But $s a = s F$ (equal tangents) and similarly $Va = Vn$; therefore $s F : s r :: Vn : VR$, which ratio is less than unity; therefore a is a point on an ellipse.*

The plane of the intersecting lines Va and Rr cuts the plane MN in AT which is therefore parallel to ar; therefore $s A : s T :: Vn : VR$. But $s A = s F_1$ therefore $s F_1 : s T :: Vn : VR$, the same ratio as before.

143. If the plane of section PN were to approach parallelism to VC the point R would advance toward n, and when VR became Vn the plane would have reached the position to give the parabola.

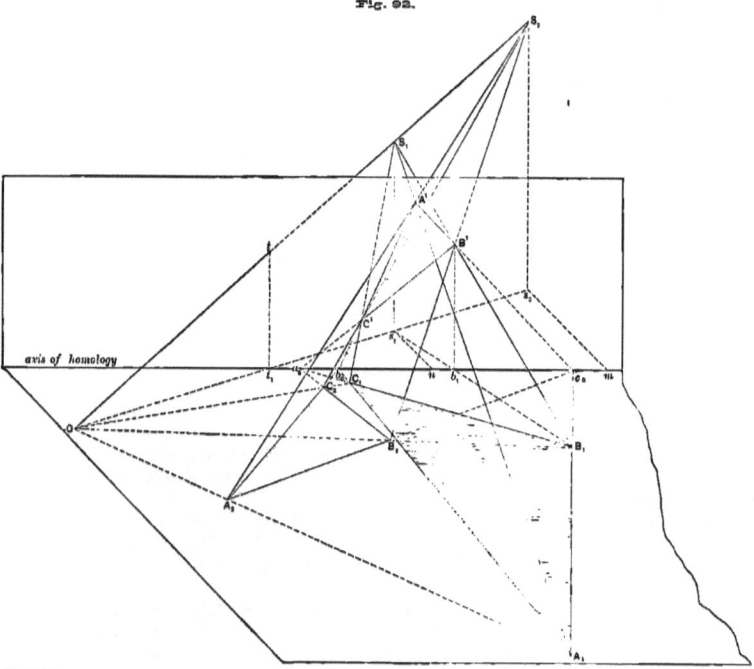

Fig. 92.

---
* Schloemilch, *Geometrie des Maasses*, 1874.

CONIC SECTIONS AS HOMOLOGOUS FIGURES.

144. The proof that KsH (Fig. 91) is an ellipse when the curve is referred to *two foci* is as follows: $KF = Km$; $KF_1 = Kt$; therefore $KF + KF_1 = Km + Kt = tm = xn = 2KF + FF_1 = 2HF + FF_1$; i. e., $KF = HF_1$.

Since $sF = sa$ and $sF_1 = sA$ we have $sF + sF_1 = sa + sA = Aa = tm = xn = HK$. The sum of the distances from any point s to the two fixed points F and F_1 is therefore constant and equal to the longer axis HK.

HOMOLOGOUS PLANE AND SPACE FIGURES.

145. Before leaving the conic sections their construction will be given by the methods of Projective Geometry. (See Art. 9.)

In Fig. 92, if S_1 is a centre of projection, then, by Art. 4, the figure $A^1 B^1 C^1$ is the *central projection* of $A_1 B_1 C_1$. The points A^1 and A_1 are *corresponding points*. Similarly B^1 and B_1, C^1 and C_1.

For S_2 as the centre of projection the figure $A_2 B_2 C_2$ corresponds to $A^1 B^1 C^1$.

If we join S_2 with S_1 and prolong to O, to meet the plane of the figures $A_1 B_1 C_1$ and $A_2 B_2 C_2$, we find that the point O sustains the same relation to these two figures that S_1 and S_2 do to the figures just taken in connection with them. Using the technical term *trace* for the intersection of a line with a plane or of one plane with another, we see that $A^1 c_0$ is the trace, on the vertical plane, of any plane containing $A^1 B^1$. This plane cuts the "axis of homology," t, m, in c_0. As $A_1 B_1$ lies in the plane of S_1 and $A^1 B^1$, and in the horizontal plane as well, it can only meet the vertical plane in c_0, the point of intersection of all these planes. Similarly we find that $A^1 C^1$ and $A_1 C_1$, if prolonged, meet the axis at the same point; correspondingly $B^1 C^1$ and $B_1 C_1$ meet at a_0. But $A^1 B^1$ and $A_1 B_1$, being corresponding lines, lie in the plane with S_1, though belonging to figures in two other planes; they must, therefore, meet also at the same point, c_0; and similarly for the other lines in the figures used with S_2.

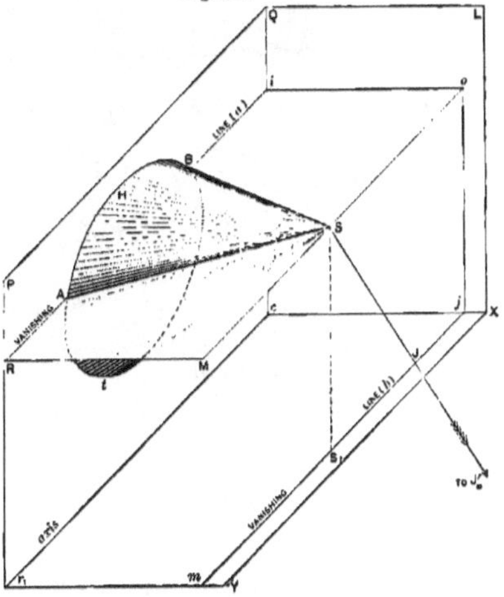

Finally, the line $A_1 A_2$ being the horizontal trace of the plane determined by the lines joining A^1 with S_1 and S_2 must contain the horizontal trace, O, of the line joining S_2 with S_1. But this puts A_2 and A_1 into the same relation with O that A_2 and A^1 sustain to S_1; or that of A^1 and A_1 to S_1. Hence we rightly conclude that *on one plane* we can take a point O as a centre of projection, and a figure $A_2 B_2 C_2$, and from them derive a second figure, $A_1 B_1 C_1$, which *corresponds* to the assumed figure in the same way as if they lay in different planes. Figures so related in one plane are called *homological figures* and the centre, O, a *centre of homology*.

146. Had $A_1 B_1 C_1$ been a *circle*, and all its points joined with S_1, then from what has preceded we know that $A^1 B^1 C^1$ would have been an ellipse; as also would have been the case were $A_2 B_2 C_2$ a circle used in connection with S_2. But our conclusions should enable us to substitute a circle for $A_2 B_2 C_2$ and using

Fig. 92.

O in the same plane with it get an ellipse in place of the triangle $A_1 B_1 C_1$. Before illustrating this it is necessary to show the relation of the axis to the other elements of the problem and to supply a test as to the nature of the conic.

147. First as to the *axis*, and employing again for a time a space figure (Fig. 93), it is evident that raising or lowering the horizontal plane $c X Y$ parallel to itself, and with it, necessarily, the axis, would not alter the *kind* of curve that it would cut from the cone S. If $A B$, were the elements of the latter prolonged. But raising or lowering the *centre*, S, would decidedly affect the curve. Where it is, there are two elements of the cone, $S A$ and $S B$, which would never meet the plane $c X Y$. The shaded plane containing those elements meets the vertical plane in "vanishing line (a)," parallel to the axis. This contains the projections, A and B_1 of the points at infinity where the lower plane may be considered as cutting the elements $S A$ and $S B$. Were S and the shaded plane raised to the level of H, so that "vanishing line (a)" should become tangent to the base, there would be one element, $S H$, of the cone, parallel to the lower plane, and the section of the cone by the latter would be the parabola; while the figure as it stands would indicate the hyperbola. The former would have but one point at infinity; the latter, two.

148. Raise the centre S so that the vanishing line does not cut the base and, evidently, no line from S to the base would be parallel to the lower plane; but the latter would cut all the elements on one side of the vertex, giving the ellipse.

149. Bearing in mind that the projection of the circle $A H B t$ is on the lower plane produced, if we wish to bring both these figures and the centre S into one plane, without destroying the relation between them, we may imagine the end plane $Q L X$ removed and rotation of the remaining system

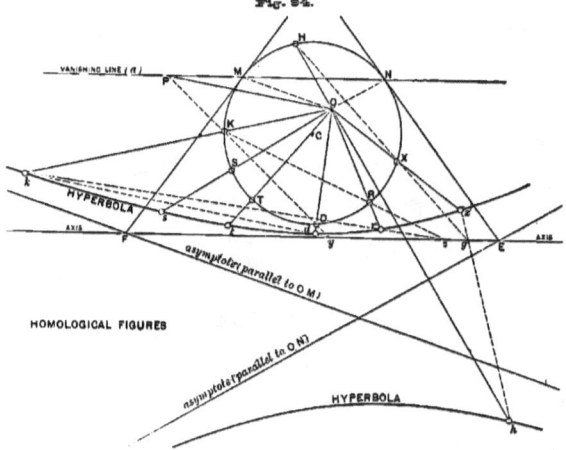

Fig. 94.

occurring about $c r_1$ in a manner exactly similar to that which would occur were $i o j c$ a system of four pivoted links and the point o pressed toward c. The motion of S would be parallel and equal

CONIC SECTIONS AS HOMOLOGOUS FIGURES.

to that of o, and, like the latter, would evidently maintain its distance from the vanishing line and describe a circular arc about it. The vanishing line would remain parallel to the axis.

150. From the foregoing we see that to obtain the hyperbola by projection of a circle from a point in the plane of the latter we would require simply a secant vanishing line, MN (Fig. 94), and an axis of homology parallel to it. Take any point P on the vanishing line and join it with any point K of the circle. PK meets the axis at y; therefore the line ky that corresponds to PK must also meet the axis at y. OP is analogous to SA of Fig. 93 in that it meets its corresponding line at infinity, i. e., is parallel to it. Therefore yk, parallel to OP meets the ray OK at k, corresponding to K. Then K joined with any other point R gives Kz. Join z with the point k just obtained and prolong OR to intersect them, obtaining r, another point of the hyperbola.

151. In Fig. 93, were a tangent at B drawn to arc AHB, it would meet the axis in a point which, like all points on the axis, corresponds to itself. From that point the projection of that tangent on the lower plane would be parallel to SB, since they are to meet at infinity. Or if SJ is parallel to the tangent at B then J will be the projection of J^1 at infinity, where SJ meets the tangent; J will be therefore one point of the projection of said tangent on the lower plane: while another point would be, as previously stated, that in which the tangent at B meets the axis.

152. Analogously in Fig. 94, the tangents at M and N meet the axis, as at F and E; but the projectors OM and ON go to points of tangency at infinity; M and N are on a "vanishing line"; hence OM is parallel to the tangent at infinity, that is to the *asymptote* (see Art. 134) through F; while the other asymptote is a parallel through E to ON.

153. As in Fig. 93 the projectors from S to all points of the arc above the level of S could cut the lower plane only by being produced to the right, giving the right-hand branch of the hyperbola; so, in Fig. 94, the arc MHN, above the vanishing line, gives the lower branch of the hyperbola. To get a point of the latter, as h, and having already obtained any point x of the other branch, join H with N (the original of x) and get its intersection, y, with the axis. Then xyh corresponds to yXH, and the ray OH meets it at h, the projection of H.

The cases should be worked out in which the vanishing line is tangent to the circle or exterior to it.

154. The homological figures with which we have been dealing were *plane* figures. But it is possible to have space figures homological with each other.

In homological space figures corresponding lines meet at the same point *in a plane*, instead of the same point on a line. A vanishing *plane* takes the place of a vanishing *line*. The figure that is in homology with the original figure is called the *relief-perspective* of the latter. (See Art. 11.)

Remarkably beautiful effects can be obtained by the construction of homological space figures, as a glance at Fig. 95 will show. The figure represents a triple row of groined arches and is from a photograph of a model designed by Prof. L. Burmester.

Fig. 95.

Although not always requiring the use of the irregular curve and therefore not strictly the material for a topic in this chapter, its close analogy to the foregoing matter may justify a few words at this point on the construction of a relief-perspective.

155. In Fig. 96 the plane PQ is called the *plane of homology* or *picture-plane*, and—adopting Cremona's notation—we will denote it by π. The *vanishing plane*, MN, or ϕ', is parallel to it. O is

50   THEORETICAL AND PRACTICAL GRAPHICS.

the *centre of homology* or *perspective-centre*. All points in the plane π are their own perspectives or, in other words, *correspond to themselves*. Therefore B'' is one point of the projection or perspective of the line AB, being the intersection of AB with π. The line Ov, parallel to AB, would meet the

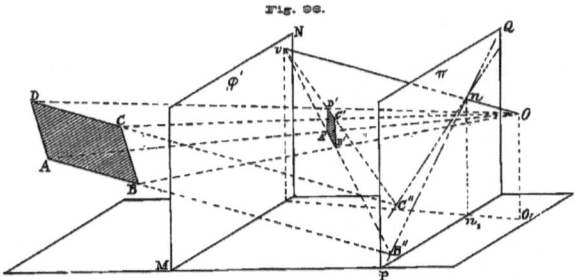

Fig. 96.

latter *at infinity;* therefore v, in the vanishing plane, φ', would be the projection upon it of the point at infinity. Joining v with B'' and cutting vB'' by rays OA and OB gives $A'B'$ as the relief-perspective of AB. The plane through O and AB cuts π in $B''n$, which is an *axis of homology* for AB and $A'B'$, exactly as mn in Fig. 92 is for $A_1 B_1$ and $A_2 B_2$.

Fig. 97.

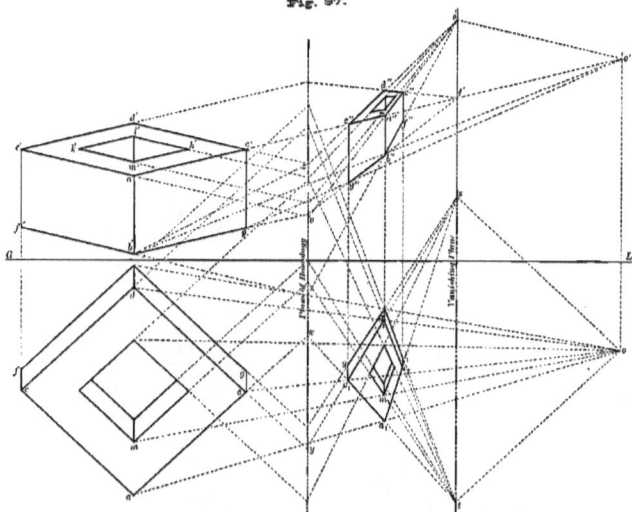

As DC is parallel to AB (Fig. 96) a parallel to it through O is again the line Ov.

The trace of DC on π is C''. Joining v with C''' and cutting vC''' by rays OD, OC, obtains $D'C''$ in the same manner as $A'B'$ was derived. The originals of $A'B'$ and $C'D'$ are parallel lines; but we see that their relief-perspectives meet at v. The vanishing plane is therefore the locus* of the vanishing points of lines that are parallel on the original object; while the plane of homology is the locus of the axes of homology of corresponding lines; or, differently stated, any line and its relief-perspective will, if produced, meet on the plane of homology.

156. Fig. 97 is inserted here for the sake of completeness, although its study may be reserved, if necessary, until the chapter on projections has been read. In it a solid object is represented at the left in the usual views, plan and elevation; GL being the *ground line* or axis of intersection of the planes on which the views are made. The planes π and ϕ' are interchanged, as compared with their positions in Fig. 96, and they are seen as lines, being assumed as perpendicular to the paper. The relief-perspective appears between them, in plan and elevation.

The lettering of AB and DC, and the lines employed in getting their relief-perspectives, being identical with the same constructions in Fig. 96 ought to make the matter clear at a glance to all who have mastered what has preceded.

Burmester's *Grundzüge der Relief-Perspective* and Wiener's *Darstellende Geometrie* are valuable reference works on this topic for those wishing to pursue its study further; but for special work in the line of homological *plane* figures the student is recommended to read Cremona's *Projective Geometry* and Graham's *Geometry of Position*, the latter of which is especially valuable to the engineer or architect since it illustrates more fully the practical application of central projection to Graphical Statics.

LINK-MOTION CURVES.

157. *Kinematics* is the science which treats of pure motion, *regardless of the cause or the results of the motion*.

It is a purely kinematic problem if we lay out on the drawing-board the path of a point on the connecting-rod of a locomotive, or of a point on the piston of an oscillating cylinder, or of any point on one of the moving pieces of a mechanism. Such problems often arise in machine design, especially in the invention or modification of valve-motions.

Some of the motion-curves or point-paths that are discovered by a study of relative motion are without special name. Others, whose mathematical properties had already been investigated and the curves dignified with names, it was later found could be mechanically traced. Among these the most familiar examples are the Ellipse and the Lemniscate, the latter of which is employed here to illustrate the general problem.

The moving pieces in a mechanism are rigid and inextensible and are always under certain *conditions of restraint*. "Conditions of restraint" may be illustrated by the familiar case of the connecting-rod of the locomotive, one end of which is always attached to the driving-wheel at the crank-pin and is therefore constrained to describe a circle about the axle of that wheel, while the other end of the rod must move in a straight line, being fastened by the "wrist-pin" to the "crosshead," which slides between straight "guides." The first step in tracing a point-path of any mechanism is therefore the determination of the *fixed* points and a general analysis of the motion,

---

* *Locus* is the Latin for *place;* and in rather unmechanical language, we may say that the *locus* of points or lines is the place where you may expect to find them under their conditions of restriction. For example, the surface of a sphere is the *locus* of all points equidistant from a fixed point (its centre). The *locus* of a point moving in a plane so as to remain at a constant distance from a given fixed point, is a circle having the latter point as its centre.

52   THEORETICAL AND PRACTICAL GRAPHICS.

158. We have given, in Fig. 98, two links or bars, MN and SP, fastened at N and P by pivots to a third link, NP, while their other extremities are pivoted on *stationary* axes at M and S. The only movement possible to the point N is therefore in a circle about M; while P is equally limited to circular motion about S. The points on the link NP, with the exception of its

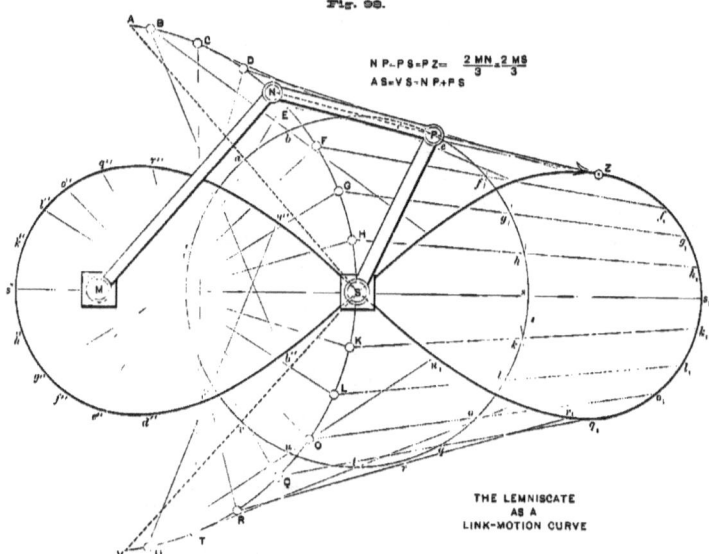

Fig. 98.

$$NP = PS = PZ = \frac{2MN}{3} = \frac{2MS}{3}$$
$$AS = VS = NP + PS$$

THE LEMNISCATE
AS A
LINK-MOTION CURVE

extremities, have a compound motion, in curves whose form it is not easy to predict and which differ most curiously from each other. The figure-of-eight curve shown, otherwise the "Lemniscate of Bernoulli," is the point-path of Z, the link NP being supposed prolonged by an amount, PZ, equal to NP. Since NP is constant in length, if N were moved along to E the point P would have to be at a distance NP from E and also on the circle to which it is confined; therefore its new position, f, is at the intersection of the circle Psr by an arc of radius PN, centre E. Then Ff prolonged by an amount equal to itself gives f_1, another point of the Lemniscate, and to which Z has then moved. All other positions are similarly found.

If the motion of N is toward D it will soon reach a limit, A, to its further movement in that direction, arriving there at the instant that P reaches a, when NP and PS will be in one straight line, SA. In this position any movement of P either side of a will drag N back over its former path; and unless P moves to the left, past a, it would also retrace its path. P reaches a similar "dead point" at v.

To get the curve illustrated, the links NP and PS had to be equal, as also the distance MS to MN. By varying the proportions of the links, the point-paths would be correspondingly affected.

INSTANTANEOUS CENTRES.—CENTROIDS. 58

By tracing the path of a point on PN produced, and as far from N as Z is from P, the student will obtain an interesting contrast to the Lemniscate.

If M and S were joined by a link and the latter held rigidly in position, it would have been called the *fixed link*; and although its use would not have altered the motions illustrated and it is not essential that it should be drawn, yet in considering a mechanism, as a whole, the line joining the fixed centres always exists, in the imagination, as a link of the complete system.

INSTANTANEOUS CENTRES.—CENTROIDS.

159. Let us imagine a boy about to hurl a stone from a sling. Just before he releases it he runs forward a few steps, as if to add a little extra impetus to the stone. While taking those few steps a peculiar shadow is cast on the road by the end of the sling, if the day is bright. The

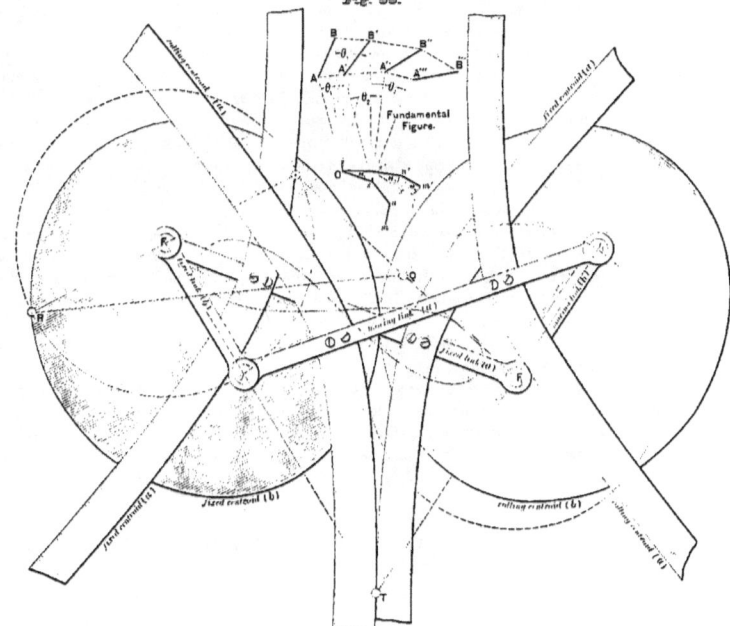

Fig. 99.

boy moves with respect to the earth; his hand moves in relation to himself, and the end of the sling describes a *circle* about his hand. The last is the only definite element of the three, yet it is sufficient to simplify otherwise difficult constructions relating to the complex curve which is described relatively to the earth.

A tangent and a normal to a *circle* are easily obtained, the former being, as need hardly be stated at this point, perpendicular to the radius at the point of tangency, while the normal simply coincides in direction with such radius. If the stone were released at any instant it would fly off in a straight line tangent to the circle it was describing about the hand as a centre; but such line would, at the instant of release, be tangent also to the compound curve. If then we wish a tangent at a given point of any curve generated by a point in motion, we have but to reduce that motion to circular motion about some moving centre; then joining the point of desired tangency with the—at that instant—position of the moving centre, we have the *normal*, a perpendicular to which gives the tangent desired.

A centre which is thus used *for an instant only* is called an *instantaneous centre*.

160. In Fig. 99 a series of instantaneous centres are shown and an important as well as interesting fact illustrated, viz., that every moving piece in a mechanism might be rigidly attached to a certain curve and by the rolling of the latter upon another curve, the link might be brought into all the positions which its visible modes of restraint compel it to take.

161. In the "Fundamental" part of the figure AB is assumed to be one position of a link. We next find it, let us suppose, at $A'B'$, A having moved over AA' and B over BB'. Bisecting AA' and BB' by perpendiculars intersecting at O, and drawing OA, OA', OB, and OB', we have $AOA' = \theta_1 = BOB'$, and O evidently a point about which, as a centre, the turning of AB through the angle θ_1 would have brought it to $A'B'$. Similarly, if the next position in which we find AB is $A''B''$, we may find a point s as the centre about which it might have turned to bring it there; the angle being θ_2, probably different from θ_1.

The points n and m are analogous to O and s.

If Os' be drawn equal to Os and making with the latter an angle θ_1, equal to the angle AOA', and if Os were rigidly attached to AB the latter would be brought over to $A'B'$ by bringing Os' into coincidence with Os. In the same manner, if we bring $s'n'$ upon sn through an angle θ_2 about s, then the next position, $A''B''$, would be reached by AB. $O's'n'm'$ is then part of a polygon whose rolling upon $Osnm$ would bring AB into all the positions shown, provided the polygon and the line were so attached as to move as one piece. Polygons whose vertices are thus obtained are called *central polygons*.

If *consecutive* centres were joined we would have curves, called *centroids*, instead of polygons; the one corresponding to $Osnm$ being called the *fixed*, the other the *rolling* centroid. The perpendicular from O upon AA' is a normal to that path. But if A were to move in a circular arc the normal to its path at any instant would be simply the radius to the position of A at that instant.

If then both A and B were moving in circular paths we would find the instantaneous centre at the intersection of the normals (radii) to the points A and B.

162. In Fig. 98 the instantaneous centre about which the whole link NP is turning is at the intersection of radii MN and SP (produced); and calling it X we would have XZ normal at Z to the Lemniscate.

163. The shaded portions of Fig. 99 illustrate some of the forms of centroids.

The mechanism is of four links, opposite links equal. Unlike the usual quadrilateral fulfilling this condition, the long sides cross, hence the name "anti-parallelogram."

The "fixed link (a)" corresponds to MS of Fig. 98 and its extremities are the centres of rotation of the short links, whose ends, f and f_1, describe the dotted circles.

For the given position T is evidently the instantaneous centre. Were a bar pivoted at T and fastened at right angles to "moving link (a)," an *infinitesimal* turning about T would move "link (a)" exactly as under the old conditions.

By taking "link (a)" in all possible positions and, for each, prolonging the radii through its extremities, the points of the fixed centroid are determined. Inverting the combination so that "moving link (a)" and its opposite are interchanged, and proceeding as before, gives the points of "rolling centroid (a)."

These centroids are branches of hyperbolas having the extremities of the long links as foci.

By holding a short link stationary, as "fixed link (b)," an elliptical fixed centroid results; "rolling centroid (b)" being obtained, as before, by inversion. The foci are again the extremities of the fixed and moving links.

Obviously the curved pieces represented as screwed to the links would not be employed in a practical construction, and they are only introduced to give a more realistic effect to the figure and possibly thereby conduce to a clearer understanding of the subject.

164. It is interesting to notice that the Lemniscate occurs here under new conditions, being traced by the middle point of "moving link (a)."

The study of kinematics is both fascinating and profitable, and it is hoped that this brief glance at the subject may create a desire on the part of the student to pursue it further in such works as Reuleaux' *Kinematics of Machinery* and Burmester's *Lehrbuch der Kinematik*.

Before leaving this topic the important fact should be stated, which now needs no argument to establish, that the instantaneous centre for any position of the moving piece, is the *point of contact* of the rolling and fixed centroids.

165. We shall have occasion to use this principle in drawing tangents and normals to the

TROCHOIDS

which are the principal *Roulettes*, or *roll-traced curves*, and which may be defined as follows:—

If in the same plane one of two circles rolls upon the other without sliding, the path of any point on the radius of the rolling circle or on the radius produced is a *trochoid*.

166. *The Cycloid.* Since a straight line may be considered a circle of *infinite radius* the above definition would include the curve traced by a point on the circumference of a locomotive wheel as it rolls along the rail, or of a carriage wheel on the road. This curve is known as a *cycloid*[*] and is shown in $T\ n\ a\ b\ c$, Fig. 100. It is the proper outline for a portion of each tooth in a certain case of gearing, viz., where one wheel has an infinite radius, that is, becomes a "rack." Were T_6 a ceiling-corner of a room, and T_{12} the diagonally opposite floor-corner, a weight would slide from T_6 to T_{12} more quickly on guides curved in cycloidal shape than if shaped to any other curve or if straight. If started at c or any other point of the curve it would reach T_{12} as soon as if started at T_6.

167. In beginning the construction of the cycloid we notice first that as $T\ 1'\ D$ rolls on the straight line $A\ B$ the arrow $D\ R\ T$ will be reversed in position (as at D, T_6) as soon as the semi-circumference $T\ 3\ D$ has had rolling contact with $A\ B$. The *tracing point* will then be at T_6, its maximum distance from $A\ B$.

When the wheel has rolled itself out once upon the rail the point T will again come in contact with the rail, as at T_{12}.

---

[*] Although the invention of the cycloid is attributed to Galileo, it is certain that the family of curves to which it belongs had been known and some of the properties of such curves investigated, nearly two thousand years before Galileo's time, if not earlier. For ancient astronomers explained the motion of the planets by supposing that each planet travels uniformly round a circle whose centre travels uniformly around another circle."—Proctor, *Geometry of Cycloids*.

The distance $T T_{12}$ evidently equals $2 \pi r$, when $r = T R$.

If the semi-circumference $T A D$ (equal to πr) be divided into any number of equal parts; and also the path of centres $R R_6$ (again $= \pi r$) into the same number of equal parts, then as the points $1, 2$, etc., come in contact with the rail, the centre R will take the positions R_1, R_2, etc., directly above the corresponding points of contact. A sufficient rolling of the wheel to bring point 2 upon $A B$ would evidently raise T from its original position to the former level of 2. But as T must always be at a radius' distance from R, and the latter would by that time be at R_2, we would find T located at the intersection (n) of the dotted *line of level* through 2 by an arc of radius $R T$, centre R_2. Similarly for other points.

The construction, summarized, involves the drawing of *lines of level* through *equidistant* points of division on a semi-circumference of the rolling circle, and their intersection by arcs of constant radius (that of the rolling circle) from centres which are the successive positions taken by the centre of the rolling circle.

It is worth while calling attention to a point occasionally overlooked by the novice, although almost self-evident, that in the position illustrated in the figure the point T drags behind the centre R until the latter reaches R_6, when it passes and goes ahead of it. From R_7 the line of level through 5 could be cut not alone at e by an arc of radius $e R_7$, but also in a second point; evidently but one of these points belongs to the cycloid, and the choice depends upon the direction of turning and the relative position of the rolling centre and the moving point. This matter requires more thought in drawing trochoidal curves in which both circles have finite radii, as will appear later.

Fig. 100.
THE CYCLOID.

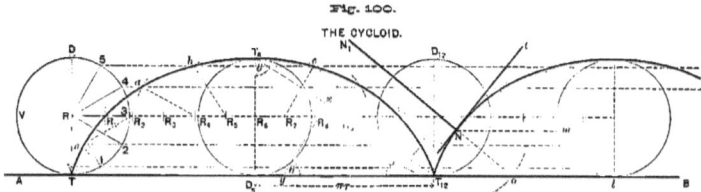

168. Were points T_0 and T_{12} given, and the semi-cycloid $T_0 T_{12}$ desired, we can readily ascertain the "base," $A B$, and generating circle, as follows: Join T_0 with T_{12}; at any point of such line, as x, erect a perpendicular, $x y$; from the similar triangles $x y T_{12}$ and $T_0 D_5 T_{12}$ having angle ϕ common and angles θ equal we see that

$$x y : x T_{12} :: T_0 D_5 : D_5 T_{12} :: 2 r : \pi r :: 2 : \pi :: 1 : \frac{\pi}{2} \text{ or, very nearly, as } 14 : 22.$$

If, then, we lay off $x T_{12}$ equal to *twenty-two* equal parts on any scale, and a perpendicular, $x y$, *fourteen* parts of the same scale, the line $y T_{12}$ will be the base of the desired curve; while the diameter of the generating circle will be the perpendicular from T_0 to $y T_{12}$ prolonged.

169. To draw a tangent to a cycloid at any point is a simple matter, if we see the analogy between the *point of contact* of the wheel and rail at any instant, and the *hand* used in the former illustration (Art. 159). At any one moment each point on the entire wheel may be considered as describing an infinitesimal arc of a circle whose radius is the line joining the point with the point of contact on the rail. The tangent at N, for example, (Fig. 100), would be $t N$, perpendicular to the normal, $N o$, joining N with o; the latter point being found by using N as a centre and

THE CYCLOID.—COMPANION TO THE CYCLOID.

cutting AB by an arc of radius equal to ml, in which m is a point *at the level* of N on any position of the rolling circle, while l is the corresponding point of contact. The point o might also have been located by the following method: Cut the line of centres by an arc, centre N, radius TR; o would obviously be vertically below the position of the rolling centre thus determined.

170. *The Companion to the Cycloid.* The kinematic method of drawing tangents, just applied, was devised by Roberval, as also the curve named by him the "Companion to the Cycloid," to which allusion has already been made (Art. 120) and which was invented by him in 1634 for the purpose of solving a problem upon which he had spent six years without success and which had foiled Galileo, viz., the calculating of the area between a cycloid and its base. Galileo was reduced to the expedient of comparing the area of the cycloid with that of the rolling circle by weighing paper models of the two figures. He concluded that the area in question was nearly but not exactly three times that of the rolling circle. That the latter would have been the correct solution may be readily shown by means of the "Companion," as will be found demonstrated in Art. 172.

171. Suppose two points coincident at T, (Fig. 101), and starting simultaneously to generate curves, the first of these points to trace the cycloid during the rolling of circle TVD while the second is to move independently of the circle and so as to be always *at the level* of the point tracing the cycloid, yet at the same time *vertically above* the point of contact of the circle and base. This makes the second point always as far from the initial vertical diameter, or *axis*, of the cycloid as the length of the arc from T to whatever level the tracing point of the latter has then reached; that is, MA equals arc THs: RO equals quadrant Tsy.

Adopting the method of Analytical Geometry, by using O as a reference point, or *origin*, we may reach any point, A, on the curve, by *co-ordinates*, as Ox, xA, of which the horizontal is called an *abscissa*, the vertical an *ordinate*. By the preceding construction Ox equals arc sfy, while xA equals sw—the *sine* of the same arc. The "Companion to the Cycloid" is therefore a *curve of sines* or *sinusoid*, since starting from O the abscissas are equal to or proportional to the arc of a circle while the ordinates are the sines of those arcs.

This curve is particularly interesting as "expressing the law of the vibration of perfectly elastic solids; of the vibratory movement of a particle acted upon by a force which varies directly as the distance from the origin; approximately, the vibratory movement of a pendulum; and exactly the law of vibration of the so-called mathematical pendulum."[*]

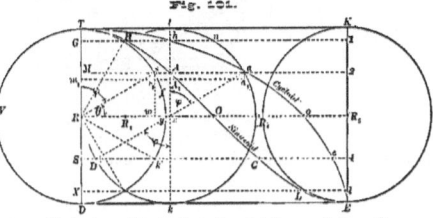

Fig. 101.

172. From the symmetry of the sinusoid with respect to RR_2 and to O we have *area* $TAOR = ECOR_2$; adding *area* $DELOR$ to both members we have the area between the sinusoid and TD and DE equal to the rectangle RE, or one-half the rectangle $DEKT$; or $\frac{1}{2} \pi r \times 2r = \pi r^2$, the area of the rolling circle.

As $TACE$ is but half of the entire sinusoid it is evident that the total area below the curve is twice that of the generating circle.

The area between the cycloid and its "companion" remains to be determined, but is readily ascertained by noting that as any point of the latter, as A, is on the vertical diameter of the circle

---
[*] Wood, *Elements of Co-ordinate Geometry*, p. 209.

passing through the then position of the tracing point, as a, the distance, $A a$, between the two curves at any level, is merely the semi-chord of the rolling circle at that level. But this, evidently, equals $M s$, the semi-chord at the same level on the equal circle. The equality of $M s$ and $A a$ makes the elementary rectangles $M s s_1 m_1$ and $A A_1 a_1 a$ equal; and considering all the possible similarly constructed rectangles of infinitesimal altitude, the sum of those on semi-chords of the rolling circle would equal the area of the semi-circle $T D y$, which is therefore the extent of the area between the two curves under consideration.

The figure showing but half of a cycloid, the total area between it and its "companion" must be that of the rolling circle. Adding this to the area between the "companion" and the base makes the total area between cycloid and base equal to *three* times that of the rolling circle.

173. The paths of points carried by and in the plane of the rolling circle, though not on its circumference, are obtained in a manner closely analogous to that employed for the cycloid.

In Fig. 102 the looped curve, traced by the arrow-point while the circle $C H M$ rolls on the base $A B$, is called the *Curtate Trochoid*. To obtain the various positions of the tracing point T describe a circle through it from centre R. On this circle lay off any even number of equal arcs and draw radii from R to the points of division; also "lines of level" through the latter. The radii drawn intercept equal arcs on the rolling circle $C H M$. While the first of these arcs rolls upon $A B$ the point T turns through the angle $T R 1$ about R and reaches the line of level through point 1. But T is always at the distance $R T$ (called the *tracing radius*) from R; and, as R has reached R_1 in the rolling supposed, we find T_1—the new position of T—by an arc from R_1, radius $T R$, cutting said line of level.

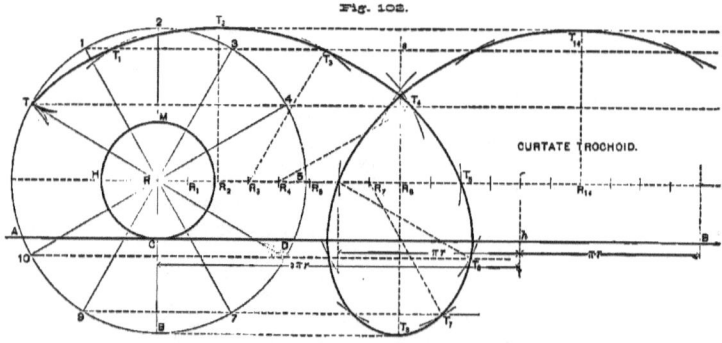

Fig. 102.

CURTATE TROCHOID.

After what has preceded, the figure may be assumed to be self-interpreting, each position of T being joined with the position of R which determined it.

174. Were a tangent wanted at any point, as T_1, we have, as before, to determine the point of contact of rolling circle and line when T reached T_1, and use it as an instantaneous centre. T_1 was obtained from R_1; and the point of contact must have been vertically below the latter and on $A B$. Joining such point to T_1 gives the normal, from which the tangent follows in the usual way.

175. *The Prolate Trochoid.* Had we taken a point *inside* of the circle $C H M$ and constructed its path the only difference between it and the curve illustrated would have been in the *name* and the *shape* of the curve. An undulating, wavy path would have resulted, called the *prolate trochoid;* but, as before, we would have described a circle through the tracing point; divided it into equal parts; drawn lines of level, and cut them by arcs of constant radius, using as centres the successive positions of *R*.

HYPO-, EPI- AND PERI-TROCHOIDS.

176. Circles of *finite* radius can evidently be tangent in but two ways—either *externally*, or *internally;* if the latter, the larger may roll on the one within it, or the smaller may roll inside the larger. When a small circle rolls within a larger the *radius* of the latter may be greater than the *diameter* of the rolling circle, or may equal it, or be smaller. On account of an interesting property of the curves traced by points in the planes of such rolling circles, viz., their capability of being generated, trochoidally, in two ways, a nomenclature was necessary which should indicate how each curve was obtained. This is included in the tabular arrangement of names below and which was the outcome of an investigation made by the writer in 1887 and presented before the American Association for the Advancement of Science.* In accepting the new terms advanced at that time Prof. Francis Reuleaux suggested the names *Ortho-cycloids* and *Cyclo-orthoids* for the classes of curves of which the cycloid and involute are respectively representative; *orthoids* being the paths of points in a fixed position with respect to a straight line rolling upon *any* curve, and *cyclo-orthoid* therefore implying a circular director or base-curve. These appropriate terms have been incorporated in the table.

For the last column a point is considered as *within* the rolling circle of infinite radius when on the normal to its initial position and on the side toward the centre of the fixed circle.

As will be seen by reference to the Appendix, the curves whose names are preceded by the same letter may be identical. Hence the terms *curtate* and *prolate*, while indicating whether the tracing point is beyond or within the circumference of the rolling circle, give no hint as to the actual *form* of the curves.

In the table *R* represents the radius of the rolling circle. *F* that of the fixed circle.

NOMENCLATURE OF TROCHOIDS.

Position of Tracing or Describing Point.	Circle rolling upon Straight Line.	External contact.	Circle rolling upon circle.					Straight Line rolling upon Circle.
			Larger Circle rolling.	Internal contact.				
				2 R > F.	2 R = F.	2 R < F.		
	Ortho-cycloids.	Epitrochoids.	Peritrochoids.	Major Hypotrochoids.	Minor Hypotrochoids.	Medial Hypotrochoids.		Cyclo-orthoids.
On circumference of rolling circle.	Cycloid.	(a) Epicycloid.	(a) Pericycloid	(d) Major Hypocycloids.	(d) Minor Hypocycloid.	Straight Hypocycloid.		Involute.
Within Circumference.	Prolate Trochoid.	(b) Prolate Epitrochoid.	(c) Prolate Peritrochoid.	(e) Major Prolate Hypotrochoid.	(f) Minor Prolate Hypotrochoid.	(g) Prolate Elliptical Hypotrochoid.		Prolate Cyclo-orthoid.
Without Circumference.	Curtate Trochoid.	(c) Curtate Epitrochoid.	(b) Curtate Peritrochoid.	(f) Major Curtate Hypotrochoid.	(e) Minor Curtate Hypotrochoid.	(g) Curtate Elliptical Hypotrochoid.		Curtate Cyclo-orthoid.

177. From the above we see that the prefix *epi* (*over* or *upon*) denotes the curves resulting from external contact; *hypo* (*under*) those of internal contact with smaller circle rolling; while *peri* (*about*) indicates the third possibility as to rolling.

* Re-printed in substance in the Appendix.

178. The construction of these curves is in closest analogy to that of the cycloid. If, for example, we desire a major hypocycloid we first draw two circles, $m\,V\,P$, $m\,x\,L$, (Fig. 103), tangent internally, of which the rolling circle has its diameter greater than the radius of the fixed circle. Then, as for the cycloid, if the *tracing-point* is P, we divide the semi-circumference $m\,V\,P$ into equal parts and from the *fixed centre*, F, describe circles through the points of division, as those through *1, 2, 3, 4* and *5*. These replace the "lines of level" of the cycloid, and may be called *circles of distance*, as they show the distance from F of the point P, for definite amounts of angular rotation of the latter. For, if the circle $P\,V\,m$ were simply to rotate about R, the point P would reach m during a semi-rotation and would then be at its maximum distance from F. After turning through the equal arcs $P\,1$, 1-2, etc., its distances from F would be $F\,a$ and $F\,b$ respectively.

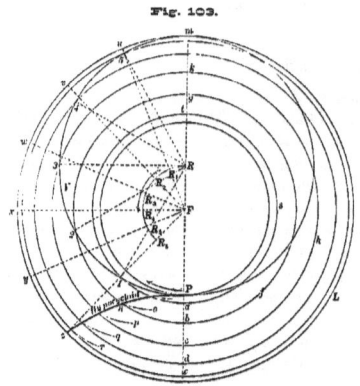

Fig. 103.

If, however, the turning of P about R is due to the rolling of circle $P\,V\,m$ upon the arc $m\,x\,z$ then the *actual position* of P, for any amount of turning about R, is determined by noting the new position of R, due to such rolling, as R_1, R_2, etc., and from it as a centre cutting the proper circle of distance by an arc of radius $R\,P$.

Since the radius of the smaller circle is in this case three-fourths that of the larger, the angle $m\,F\,z$ (135°) at the centre of the latter intercepts an arc, $m\,x\,z$, *equal to the* 180° arc, $m\,V\,P$, on the smaller circle. *Equal arcs on unequal circles are subtended by angles at the centre which are inversely proportional to the radii.*

While arc $m\,V\,P$ rolls upon arc $m\,x\,z$ the centre R will evidently move over circular arc $R \cdots R_6$. Divide $m\,x\,z$ into as many equal parts as $m\,V\,P$ and draw radii from F to the points of division; these cut the path of centres at the successive positions of R. When arc $m\,5$-4, for example, has rolled upon its equal $m\,u\,v$ then R will have reached R_2; P will have turned about R through angle $P\,R\,2 = m\,R\,4$ and will be at n, the intersection of $b\,f\,g$—the circle of distance through 2—by an arc, centre R_2, radius $R\,P$. Similarly for other points.

179. To trace the path of any point on the circumference of a circle so rolling as to give the epi- or peri-cycloid requires a construction similar at every step to that just illustrated. The same remark applies equally to the determination of the paths of points within or beyond the circumference of the rolling circle, as will be seen by reference to Fig. 104, in which the path of the point P is determined (a) as carried by the circle called "first generator," rolling on the exterior of the "first director"; (b) as carried by the "second generator" which rolls on the exterior of the "second director"—which it also encloses. In the first case the resulting curve is a *prolate epitrochoid*; in the second a *curtate peritrochoid*. Proceeding in the usual way, a semi-circle is drawn through P from each rolling centre, R and ρ. Dividing these semi-circles into the same number of equal parts draw next the dotted "circles of distance" through these points, all from centre F. The figure illustrates the special case where the larger set of "circles of distance" divides both semi-circumferences into equal parts. The successive positions of P, as P_1, P_2, etc., are then located by

EPI- AND PERI-TROCHOIDS.

arcs of radii RP or ρP_1 struck from the successive positions of R or ρ and intersecting the proper "circle of distance." For example, the turning of P through the angle $PR1$ about R would bring P somewhere upon the circle of distance through point 1; but that amount of turning would be due to the rolling of the first generator over the arc mQ_1 which would carry R to R_1; P would

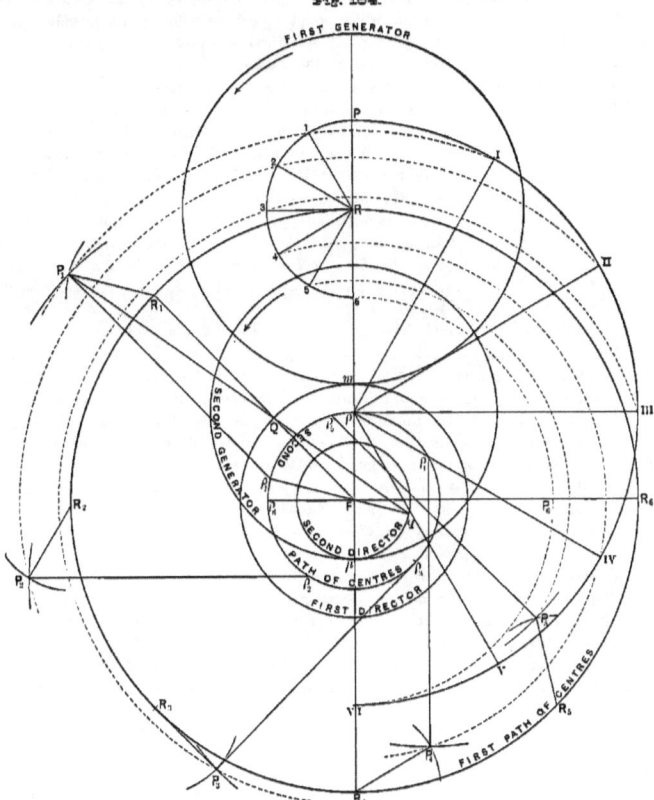

Fig. 104.

therefore be at P_1, at a distance RP from R_1 and on the dotted arc through 1. Similarly in relation to ρ. Each position of P is joined with each of the centres from which it could be obtained.

SPECIAL TROCHOIDS.

180. *The Ellipse and Straight Line.* Two circles are called Cardanic* if tangent *internally* and the diameter of one is twice that of the other. If the smaller roll in the larger all points in the plane of the generator will describe *ellipses* except points on the circumference, each of which will move in a straight line—a diameter of the director. Upon this latter property the mechanism known as "White's Parallel Motion" is based, in which a piston-rod or other piece intended to have reciprocating rectilinear motion is pivoted to a small gear-wheel or pinion which rolls on the interior of a toothed annular wheel of twice the diameter of the pinion.

181. *The Limaçon and Cardioid.* The Limaçon is a curve whose points may be obtained by drawing random secants through a point on the circumference of a circle and on each laying off a constant distance, on each side of the second point in which the secant cuts the circle.

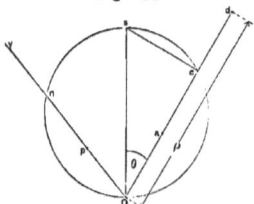

Fig. 105.

In Fig. 105 let Oc and Od be random secants of the circle Ons; then if nr, np, ca and cd are each equal to some constant, b, we shall have r, p, a and d as four points of a Limaçon. Referring any point as d to O and the diameter Os, we have $Od = \rho = Oc + cd = 2r\cos\theta + b$, while $Oa = 2r\cos\theta - b$, whence the general polar equation for the Limaçon, $\rho = 2r\cos\theta + b$.

When $b = 2r$ the Limaçon becomes the heart-shaped curve called the *Cardioid.*†

182. All Limaçons, general and special, may be generated either as epi- or peri-trochoidal curves: as epi-trochoids the generator and director must have *equal* diameters, any point on the circumference of the generator then tracing a Cardioid, while any point on the radius (or radius produced) describes a Limaçon; as peri-trochoids the larger of a pair of Cardanic circles must roll on the smaller, the Cardioid and Limaçon then resulting, as before, from the motion of points respectively *on* the circumference of the generator or *within* or *without* it.

183. In Fig. 106 the Cardioid is obtained as an epicycloid, being traced by point P during one revolution of the generator PHm about an equal directing circle msO.

As a Limaçon we may get points of the Cardioid, as y and z, by drawing a secant through O and laying off sy and sz each equal to $2r$.

184. *The Limaçon as a Trisectrix.* Three famous problems of the ancients were the squaring of the circle, the duplication of the cube and the trisection of an angle. Among the interesting curves invented by early mathematicians for the purpose of solving the latter problem were the Quadratrix and Conchoid, whose construction is given later in this chapter; but it has been found that certain trochoids also possess this interesting property, among them the Limaçon of Fig. 106, frequently called the *Epitrochoidal Trisectrix.* Its construction as an epitrochoid need not be described in detail, after what has preceded.

As a Limaçon we would find points as G and X by drawing from R a secant RX to the circle called "path of centres" and making SX and SG each equal to the radius of that circle.

185. To trisect an angle, as MRF, bisect one side of the angle as RF at m; use mR and mF as radii for generator and director respectively of an epitrochoid having a tracing radius, RF, equal to twice that of the generator. Make $RN = RF$ and draw NP; this will cut the Limaçon

---

* Term due to Reuleaux and based upon the fact that Cardano (16th century) was probably the first to investigate the paths described by points during their rolling.
† From *Cardis*, the Latin for *heart*.

SPECIAL TROCHOIDS.

$F T_1 R Q$ (traced by point F as carried by the given generator) in a point T_1. The angle $T_1 R F$ will then be one-third of $N R F$, which may be proved as follows: F reaches T_1 by the rolling of arc $m n$ on arc $m n_1$. These arcs are subtended by equal angles, ϕ, the circles being equal. During this rolling R reaches R_1, bringing $R F$ to $R_1 T_1$. In the triangles $T_1 R_1 F$ and $R F R_1$ the side $F R_1$ is common, angles ϕ equal and side $R_1 T_1$ equal to side $R F$; the line $R T_1$ is therefore parallel to $R_1 F$, whence angle $T_1 R F$ must also equal ϕ. In the triangle $R F R_1$, we denote by θ the angles opposite the equal sides $R F$ and $R_1 F$; then $2\theta + \phi = 180°$ or $\theta = \dfrac{180° - \phi}{2}$. In triangle $N R F$ we have the angle at F equal to $\theta - \phi$, and $2(\theta - \phi) + x + \phi = 180°$ which gives $x = 2\phi$ by substituting value of θ from previous equation.

Fig. 108.

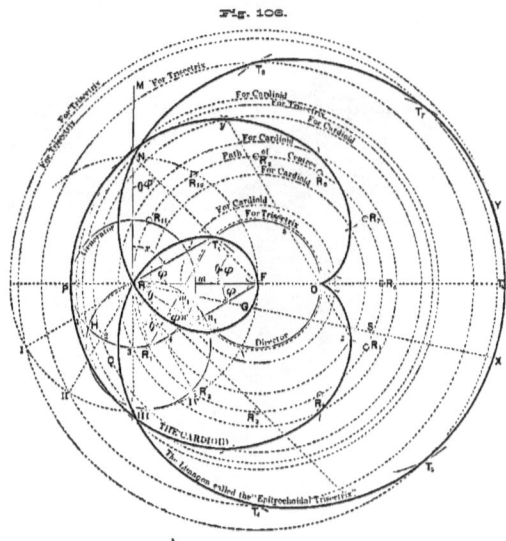

186. The Involute. As the opposite extreme of a circle rolling on a straight line we may have the latter rolling on a circle. In this case the *rolling* circle has an *infinite radius*. A point on the straight line describes a curve called the *involute*. This would be the path of the end of a thread if the latter were in tension while being unwound from a spool.

In Fig. 107 a rule is shown, tangent at u to a circle on which it is supposed to roll. Were a pencil-point inserted in the centre of the circle at j (which is on line $u x$ produced) it would trace the involute. When j reaches a the rule will have had rolling contact with the base circle over an arc $u t s \cdots a$ whose length equals line $u x j$. Were a the initial point we may obtain b, c, etc., by making *tangent* $m b =$ *arc* $m a$; *tangent* $n c =$ *arc* $n a$, etc. Each tangent thus equals the arc from initial point to point of tangency.

187. The circle from which the involute is derived or *evolved* is called the *evolute*. Were a hexagon or other figure to be taken as an evolute a corresponding involute could be derived; but the name "involute," unqualified, is understood to be that obtained from a circle.

From the law of formation of the involute the rolling line is in all its positions a normal to the curve; the point of tangency on the evolute is an instantaneous centre, and a tangent at any point as f is a perpendicular to the tangent, fq, from f to the base circle.

Like the cycloid, the involute is a correct working outline for the teeth of gear-wheels; and gears manufactured on the involute system are to a considerable degree supplanting other forms.

A surface known as the *developable helicoid* is formed by moving a line so as to be always *tangent* to a given helix. It is interesting in this connection to notice that any plane perpendicular to the axis of the helix would cut such a surface in a pair of involutes.*

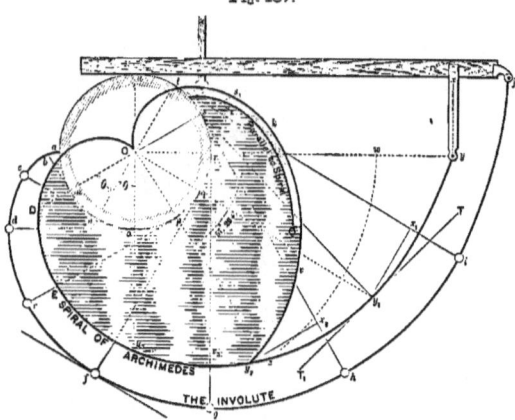

Fig. 107.

188. *The Spiral of Archimedes.* This is the curve that would be generated by a point having a uniform motion around a fixed point—the *pole*—combined with uniform motion toward or from the pole.

In Fig. 107, with O as the pole, if the angles θ are equal and OD, OE and Oy_2 are in arithmetical progression then the points D, E and y_2 are points of an Archimedean Spiral.

This spiral can be trochoidally generated, simultaneously with the involute, by inserting a pencil point at y in a piece carried by—and at right angles with—the rule, the point y being at a distance, xy, from the contact-edge of the rule, equal to the radius Os of the base circle of the involute; for

---

* The day of writing the above article the following item appeared in the New York *Evening Post*: "Visitors to the Royal Observatory, Greenwich, will hereafter miss the great cylindrical structure which has for a quarter century and more covered the largest telescope possessed by the Observatory. Notwithstanding its size the Astronomer Royal has now procured through the Lords Commissioners a telescope more than twice as large as the old one.... The optical peculiarities embodied in the new instrument will render it one of the three most powerful telescopes at present in existence.... The peculiar architectural feature of the building which is to shelter the new telescope is that its dome, of thirty-six feet diameter, will surmount a tower having a diameter of only thirty-one feet. Technically the form adopted is the surface generated by the revolution of an *involute of a circle*."

SPECIAL TROCHOIDS.

after the rolling of $a s$ over an arc $n t$ we shall have $t x_1$ as the portion of the rolling line between x and the point of tangency, and $x y$ will have reached $x_1 y_1$. If the rolling be continued y will evidently reach O. We see that $O y = a x$ and $O y_1 = t x_1$; but the lengths $a x$ and $t x_1$ are proportional to the angular movement of the rolling line about O, and as the spiral may be defined as that curve in which the length of a radius vector is directly proportional to the angle through which it has turned about the pole, the various positions of y are evidently points of such a curve.

189. Were the pole, O, given and a portion only of the spiral, we could draw a tangent at any point, y_1, by determining the circle on which the spiral could be trochoidally generated, then the instantaneous centre for the given position of the tracing-point, whence the normal and tangent would be derived in the usual way. The radius $O t$ of the base circle would equal πy_1,—the difference between two radii vectores $O y$ and $O z$ which include an angle of $57° 29 +$, (the angle which at the centre of a circle subtends an arc equal to the radius). The instantaneous centre, t, would be the extremity of that radius which was perpendicular to $O y_1$. The normal would be $t y_1$, and the tangent $T T_1$ perpendicular to it.

190. This spiral is the proper outline for a cam to convert uniform rotary into uniform rectilinear motion, and when combined with an equal and opposite spiral gives the well known form called the *heart-cam*. As usually constructed the acting curve is not the true spiral but a curve whose points are at a constant distance from the theoretical outline equal to the radius of the friction-roller which is on the end of the piece to be raised. A small portion of such a "parallel curve" is indicated in the upper part of Fig. 107.

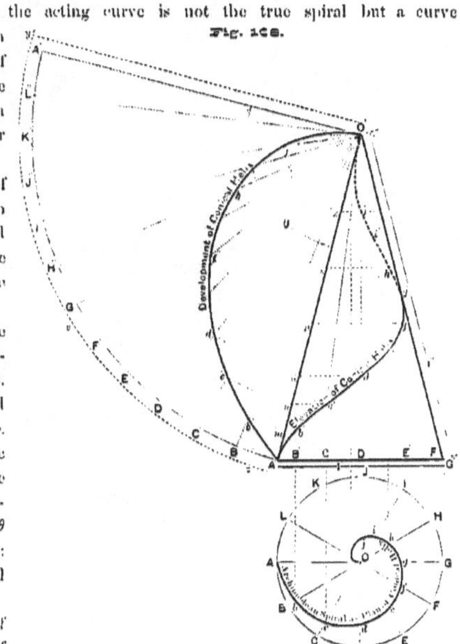

191. If a point travel on the surface of a cone so as to combine a uniform motion around the axis with a uniform motion toward the vertex it will trace a *conical helix* whose orthographic projection on the plane of the base will be a Spiral of Archimedes.

In Fig. 108 a top and front view are given of a cone and helix. The shaded portion is the *development* of the cone, that is, the area equal to the convex surface and which—if rolled up—would form the cone. To obtain the development draw an arc $A' G'' A''$ of radius equal to an element. The convex surface of the cone will then be represented by the sector $A' O' A''$ whose angle θ may be found by the proportion $O A : O' A' :: \theta : 360°$, since the arc $A' G'' A''$ must equal the entire circumference of the cone's base.

The student can make a paper model of the cone and helix by cutting out a sector of a circle, making allowance for an overlap on which to put the mucilage, as shown by the dotted lines $O' y$ and $y r z$ in the figure.

Fig. 108.

66   THEORETICAL AND PRACTICAL GRAPHICS.

The *development* of the conical helix is evidently the same kind of spiral as its *orthographic projection.*

PARALLEL CURVES.

192. A *parallel curve* is one whose points are at a constant normal distance from some other curve. Parallel curves have not the same mathematical properties as those from which they are derived, except in the case of a circle; this can readily be seen from the cam figure under the last heading, in which a point, as S_i, of the true spiral, is located on a line from O which is by no means in the direction of the *normal* to the curve at S_i, upon which lies the point S_i of the parallel curve.

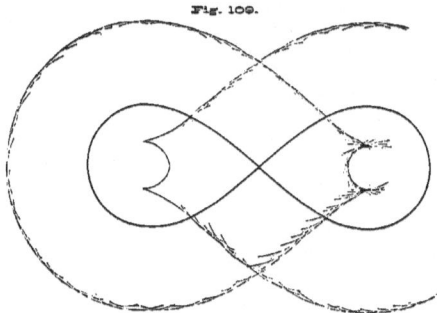

Fig. 109.

Instead of actually determining the normals to a curve and on each laying off a constant distance we may draw many arcs of constant radius, having their centres on the original curve; the desired parallel will be tangent to all these arcs.

In strictly mathematical language the parallel curve is the *envelope* of a circle of constant radius whose centre is on the original curve. We may also define it as the locus of consecutive intersections of a system of equal circles having their centres on the original curve.

If on the *convex* side of the original the parallel will resemble it in form, but if *within* the two may be totally dissimilar. This is well illustrated by Fig. 109 in which the parallel to a Lemniscate is shown.

The student will obtain some interesting results by constructing the parallels to ellipses, parabolas and other plane curves.

THE CONCHOID OF NICOMEDES.

193. The *Conchoid*, named after the Greek word for *shell*,[*] may be obtained by laying off a constant length on each side of a given line MN (the *directrix*) upon radials through a fixed point or pole, O (Fig. 110). If $mv = mn = sx$ then v, n and x are points of the curve. Denote by a the distance of O from MN and use c for the constant length to be laid off; then if $a < c$ there will be a loop in that branch of the curve which is nearest the pole,—the *inferior* branch. If $a = c$ the curve has a point or *cusp* at the pole. When $a > c$ the curve has an undulation or wave-form towards the pole.

---
[*] A series of curves much more closely resembling those of a shell can be obtained by tracing the paths of points on the piston-rod of an oscillating cylinder. See Arts. 137 and 158.

$O r = c + O m$; $O n = c — O m$; we may therefore express the relation to O of points on the curve by the equation $\rho = c \pm O m = c \pm a \sec \phi$.

Fig. 110.

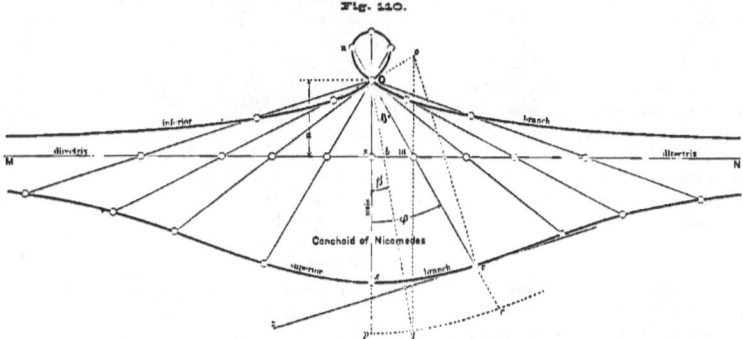

Conchoid of Nicomedes

194. Mention has already been made of the fact that this was one of the curves invented for the purpose of solving the problem of the trisection of an angle. Were the angle $m O x$ (or ϕ) we would first draw $p q c$, the superior branch of a conchoid having the constant, c, equal to twice $O m$. A parallel from m to the axis will intersect the curve at q; the angle $p O q$ will then be one-third of ϕ: for since $b q = 2 O m$ we have $m q = 2 O m \cos \beta$; also $m q : O m :: \sin \theta : \sin \beta$ (the sides of a triangle being proportional to the sines of the opposite angles); therefore $2 O m \cos \beta : O m :: \sin \theta : \sin \beta$, whence $\sin \theta = 2 \sin \beta \cos \beta = \sin 2 \beta$ (from known trigonometric relations). The angle ϕ is therefore equal to twice β, making the latter one-third of angle ϕ.

195. To draw a tangent and normal at any point r we find the instantaneous centre o on the principle that it is at the intersection of normals to the paths of two moving points of a line, the distance between said points remaining constant. The motion of r in tracing the curve is—at the instant considered—in the direction $O r$; $O o$ is therefore the normal. The point m of $O r$ is at the same moment moving along $M N$, for which $m o$ is the normal. Their intersection o is then the instantaneous centre and $o r$ the normal to the conchoid, with $r z$ perpendicular to $o r$ for the desired tangent.

196. This interesting curve may be obtained as a plane section of one of the higher mathematical surfaces. If two non-intersecting lines—one vertical, the other horizontal—be taken as guiding lines or *directrices* of the motion of a third straight line whose inclination to a horizontal plane is to be constant, then every horizontal plane will cut *conchoids* from the surface thus generated, while every plane parallel to the directrices will cut *hyperbolas*. From the nature of its plane sections this surface is called the *Conchoidal Hyperboloid*.

THE QUADRATRIX OF DINOSTRATUS.

197. In Fig. 111 let the radius $O T$ rotate uniformly about the centre; simultaneously with its movement let $M N$ have a uniform motion parallel to itself, reaching $A B$ at the same time with radius $O T$; the locus of the intersection of $M N$ with the radius will be the *Quadratrix*. Points

exterior to the circle may be found by prolonging the radii while moving MN away from AB. As the intersection of MN with OB is at infinity the former becomes an asymptote to the curve as often as it moves from the centre an additional amount equal to the diameter of the circle; the number of branches of the Quadratrix may therefore be infinite. It may be proved analytically that the curve crosses OA at a distance from O equal to $2r \div \pi$.

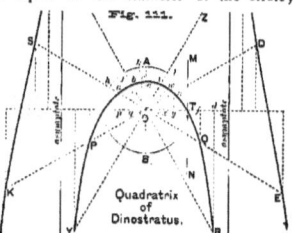

Fig. 111.

198. *To trisect an angle*, as TOa, by means of the Quadratrix draw the ordinate ap, trisect pT by s and x and draw sc and xm; radii Oc and Om will then divide the angle as desired: for by the conditions of generation of the curve the line MN takes three equidistant parallel positions while the radius describes three equal angles.

THE CISSOID OF DIOCLES.

199. This curve was devised for the purpose of obtaining two mean proportionals between two given quantities, by means of which one of the great problems of the Greek geometers—the duplication of the cube—might be effected.

The name was suggested by the Greek word for *ivy* since "the curve appears to mount along its asymptote in the same manner as that parasite plant climbs on the tall trunk of the pine."*

This was one of the first curves invented after the discovery of the conic sections. Let C (Fig. 112) be the centre of a circle, ACE a right angle, NS and MT any pair of ordinates parallel to

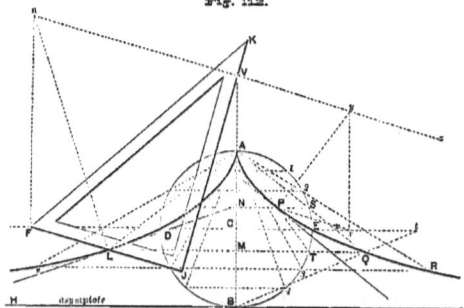

Fig. 112.

and equidistant from CE; then either secant from A through the extremity of one ordinate will meet the other ordinate in a point of the cissoid; P and Q, for example, will be points of the curve.

The tangent to the circle at B will be an asymptote to the curve.

It is a somewhat interesting coincidence that the area between the cissoid and its asymptote is the same as that between a cycloid and its base, viz., three times that of the circle from which it is derived.

---
* Leslie. *Geometrical Analysis.* 1821.

THE CISSOID.—THE TRACTRIX.

200. Sir Isaac Newton devised the following method of obtaining a cissoid by continuous motion: Make $AV = AC$; then move a right-angled triangle, of base $= VC$, so that the vertex F travels along the line DE while the edge JK always passes through V; then the middle point, I, of the base FJ, will trace a cissoid. This construction enables us readily to get the instantaneous centre and a tangent and normal; for Fn is normal to FC—the path of F, while nz is normal to the motion of J toward JV; the instantaneous centre n is therefore at the intersection of these normals. For any other point as P we apply the same principle thus: With radius equal to AC and from centre P obtain x; draw Px, then Vz parallel to it; a vertical from x will meet Vz at the instantaneous centre y, from which the normal and tangent result in the usual way.

201. Two quantities m and n will be mean proportionals between two other quantities a and b if $m^2 = na$ and $n^2 = mb$; that is, if $m^3 = a^2b$ and if $n^3 = ab^2$.

If $b = 2a$ we will have m for the edge of a cube whose volume will be twice that of a^3, when $m^3 = a^2b$.

To get two mean proportionals between quantities r and b make the smaller, r, the radius of a circle from which derive a cissoid. Were APB the derived curve we would then make Ct equal to the second quantity, b, and draw Bt, cutting the cissoid at Q. A line AQ would cut off on Ct a distance Cr equal to m, one of the desired proportionals; for m^3 will then equal r^2b, as may be thus shown by means of similar triangles:—

$$Cr : MQ :: CA : MA \text{ whence } Cc^3 = \frac{r^3 \cdot MQ^3}{MA^3} \quad . \quad (1)$$

$$Ct : MQ :: CB : BM \quad \text{``} \quad Ct = \frac{r \cdot MQ}{BM} \quad . \quad (2)$$

$$MQ : MA :: SN : AN :: \sqrt{AN \cdot BN} : AN, \text{ whence } MQ = \frac{MA\sqrt{AN \cdot BN}}{AN} \quad (3)$$

From (2) we have $MQ = \frac{BM(Ct = b)}{r}$ . . . . (4)

`` (3) `` $MQ^2 = \frac{MA^2(AN \cdot BN)}{AN^2}$ . . . . . . (5)

Replacing MQ^3 in equation (1) by the product of the second members of equations (4) and (5) gives Cr^3 (i. e., m^3) $= r^2 b$.

By interchanging r and b we obtain n, the other mean proportional; or it might be obtained by constructing similar triangles having a, b and m for sides.

THE TRACTRIX.

202. The *Tractrix* is the involute of the curve called the *Catenary* (later described) yet its usual construction is based on the fact that if a series of tangents be drawn to the curve the portions of such tangents between the points of tangency and a given line will be of the same length; or, in other words, the intercept on the tangent between the directrix and the curve will be constant. A practical and very close approximation to the theoretical curve is obtained by taking a radius QR (Fig. 113) and with a centre a, a short distance from R on QR, obtaining b which is then joined with a. On ab a centre c is similarly taken, whence cd is obtained. A sufficient repetition of this process will indicate the curve by its enveloping tangents; or a curve may actually be drawn tangent to all these lines. Could we take a, b, c, etc., as mathematically *consecutive points* the curve would be theoretically exact.

The line QS is an asymptote to the curve.

The area between the completed branch RPS and the lines QR and QS would be equal to a quadrant of the circle on radius QR.

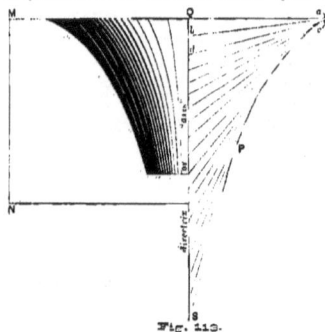

Fig. 113.

203. The surface generated by revolving the tractrix about its asymptote has been employed for the foot of a vertical spindle or shaft and is known as Schiele's Anti-Friction Pivot. The step for such a pivot is shown in sectional view in the left half of the figure. Theoretically the amount of work done in overcoming friction is the same on all equal areas of this surface.

In the case of a bearing of the usual kind, for a cylindrical spindle, although the *pressure* on each square inch of surface would be constant yet as unit areas at different distances from the centre would pass over very different amounts of space in one revolution, the wear upon them would be necessarily unequal. The *rationale* of the tractrix form will become evident from the following consideration:—If about to split a log, and having a choice of wedges, any boy would choose a thin one rather than one with a large angle, although he might not be able to prove by graphical statics the exact amount of advantage the one would have over the other. The theory is very simple, however, and the student may profitably be introduced to it. Suppose a ball, c, (Fig. 114) struck at the same instant by two others, a and b, moving at rates of six and eight feet a second respectively. On ac and bc prolonged take cc and ch equal, respectively, to *six* and *eight* units of some scale; complete the parallelogram having these lines

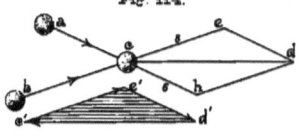

Fig. 114.

as sides; then it is a well known principle in mechanics* that cd—the diagonal of this parallelogram—will not only represent the *direction* in which the ball c will move but also the *distance*—in feet, to the scale chosen—it will travel in one second. Obviously, then, to balance the effect of balls a and b upon c, a fourth would be necessary, moving from d toward c and traversing dc in the same second that a and b travel, so that impact of all would occur simultaneously. These forces would be represented in direction and magnitude (to some scale) by the shaded triangle $c'd'e'$, which illustrates the very important theorem that if the three sides of a triangle—taken like $c'e'$, $e'd'$, $d'c'$, in such order as to bring one back to the initial vertex mentioned—represent in *magnitude* and *direction* three forces acting on one point, then these forces are balanced.

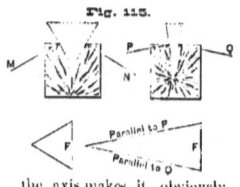

Fig. 115.

Constructing now a triangle of forces for a broad and thin wedge, (Fig. 115) and denoting the force of the supposed *equal blows* by F in each triangle, we see that the pressures are greater for the thin wedge than for the other; that is, the less the inclination to the vertical the greater the pressure. A pivot so shaped that as the *pressure* between it and its step *increased* the *area* to be traversed *diminished* would therefore, theoretically, be the ideal; and the rate of change of curvature of the tractrix as its generating point approaches the axis makes it, obviously, the correct form.

---

* For a demonstration the student may refer to Rankine's *Applied Mechanics*, Art. 51.

204. Navigator's charts are usually made by *Mercator's projection* (so-called, not being a *projection* in the ordinary sense, but with the extended signification alluded to in the remark in Art. 2). Maps thus constructed have this advantageous feature. that *rhumb lines* or *loxodromies*—the curves on a sphere that cut all meridians at the same angle—are represented as *straight lines*, which can only be the case if the meridians are indicated by *parallel* lines. The law of convergence of meridians on a sphere is, that the length of a degree of longitude at any latitude equals that of a degree on the equator multiplied by the cosine (see footnote, p. 31,) of the latitude; when the meridians are made *non-convergent* it is, therefore, manifestly necessary that the distance apart of originally equidistant parallels of latitude must increase at the same rate; or, otherwise stated, as on Mercator's chart degrees of longitude are all made equal, regardless of the latitude, the constant length representative of such degree bears a varying ratio to the actual arc on the sphere, being greater with the increase in latitude; but the greater the latitude the less its cosine or the greater its secant; hence lengths representative of degrees of latitude will increase with the secant of the latitude. Tables have been constructed giving the increments of the secant for each minute of latitude; but it is an interesting fact that they may be derived from the Tractrix thus: Draw a circle with radius QR, centre Q (Fig. 113); estimate latitude on such circle from R upward; the intercept on QS between consecutive tangents to the Tractrix will be the increment for the arc of latitude included between parallels to QS, drawn through the points of contact of said pair of tangents.*

On the subject of map construction the student is referred to Craig's *Treatise on Projections.*

THE WITCH OF AGNISI.

205. If on any line SQ, perpendicular to the diameter of a circle, a point S be so located that $SQ:AB::PQ:QB$ then S will be a point of the curve called the *Witch of Agnisi*. Such point is evidently on the ordinate PQ prolonged, and vertically below the intersection T of the tangent at A by the secant through P.

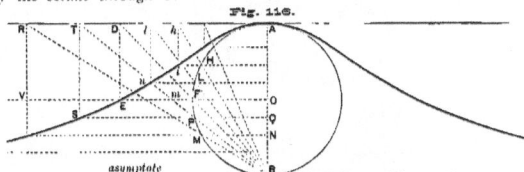

Fig. 116.

The point E, at the same level as the centre O, is a diameter's distance from the latter.

The tangent at B is an asymptote to the curve.

The area between the curve and its asymptote is four times that of the circle involved in its construction.

The *Witch*, also called the *Versiera*, was devised by Donna Maria Gaetana Agnisi, a brilliant Italian lady, who was appointed (1750) by Pope Benedict XIV. to the professorship of mathematics and philosophy in the University of Bologna.

THE CARTESIAN OVAL.

206. This curve, also called simply a *Cartesian*, after its investigator, Descartes, has its points connected with two foci, F'' and F''', by the relation $m\rho' \pm n\rho'' = kc$, in which c is the distance between the foci while m, n and k are constant factors.

---
* Leslie, *Geometrical Analysis*, Edinburgh, 1821.

72   THEORETICAL AND PRACTICAL GRAPHICS.

Fig. 117.

Salmon states that we owe to Chasles the proof that a third focus may be found, sustaining the same relation, expressed by an equation of similar form. (See Art. 209.)

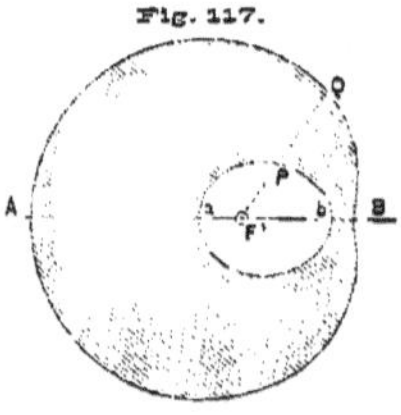

The Cartesian is symmetrical with respect to the *axis*—the line joining the foci.

207. To construct the curve from the first equation we may for convenience write $m \rho' \pm n \rho'' = k c$ in the form $\rho' \pm \frac{n}{m} \rho'' = \frac{k c}{m}$; or by denoting $\frac{n}{m}$ by b and $\frac{k c}{m}$ by d it takes the yet more simple form $\rho' \pm b \rho'' = d$. Then ρ'' will have two values according as the positive or negative sign is taken, being respectively $\frac{d-\rho'}{b}$ and $\frac{\rho'-d}{b}$; the former is for points on the *inner* of the *two* ovals that constitute a complete Cartesian, while the latter gives points on the *outer* curve.

To obtain $\rho'' = \frac{d-\rho'}{b}$ take F' and F''' (Fig. 118) as foci; $F''S = d$; SK at some random acute angle θ with the axis, and make $SH = \frac{d}{b}$; that is, make $F''S : SH :: b : 1$. Then from F' draw an arc tfP, of radius less than d, and cut it at P by an arc from centre F''', radius ST. Tt being a parallel to $F''H$; then P is a point of the inner oval; for $St = d - \rho'$ and $ST = \rho''$; therefore $\rho'' : d - \rho' :: \frac{d}{b} : d$ whence $\rho'' = \frac{d-\rho'}{b}$.

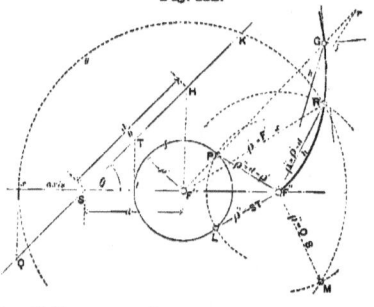

Fig. 118.

208. If an arc syk be drawn from F' with radius $F''s$, *greater than* d, we may find the second value of ρ'', viz., $\frac{\rho'-d}{b}$ by drawing sQ parallel to $F''H$ to meet HS prolonged; for QS will equal $\frac{\rho'-d}{b}$, in which $\rho' = F''s$. Again using F'' as a centre, and a radius $QS = \rho''$, gives points R and M of the larger oval.

The following are the values for the focal radii to the four points where the ovals cut the axes. (See Fig. 117.)

For A, $\rho'' = \frac{\rho'-d}{b} = c + \rho'$ whence $\rho' = F'A = \frac{d+bc}{1-b}$

" a, $\rho'' = \frac{d-\rho'}{b} = c + \rho'$ " $\rho' = F'a = \frac{d-bc}{1+b}$

B, $\rho'' = \frac{\rho'-d}{b} = c - \rho'$ " $\rho' = F'B = \frac{d+bc}{1+b}$

" b, $\rho'' = \frac{d-\rho'}{b} = c - \rho'$ " $\rho' = F'b = \frac{d-bc}{1-b}$

The construction-arcs for the outer oval must evidently have radii *between* the values of ρ' for A and B above; and for the inner oval *between* those of a and b.

The numerical values from which Fig. 118 was constructed were $m = 3$; $n = 2$; $c = 1$; $k = 3$.

209. The Third Focus. Figure 118 illustrates a special case, but in general the method of finding a third focus. F'''' (not shown), would be to draw a random secant $F'' r$ through F' and note the points P and G in which it cuts the ovals—these to be taken on the same side of F', as two other points of intersection are possible: a circle through P, G and F''' would cut the axis in the new focus sought. Then denoting by C the distance $F'' F''''$, we would find the factors of the original equation appearing in a new order. thus, $k\rho' \pm n\rho''' = m C$, which—for purposes of construction—may be written $\rho' \pm b'\rho''' = d'$.

If obtained from the foci F''' and F'''' the relation would be $m\rho''' - k\rho'' = \pm n C'$, in which C' equals $F''' F''''$. Writing this in the form $\rho''' - B\rho'' = \pm D$ we have the following interesting cases: (a) an *ellipse* for D positive and $B = -1$; (b) an *hyperbola* for D positive and $B = +1$; (c) a *limaçon* for $D = C' B$; (d) a *cardioid* for $B = +1$ and $D = C'$.

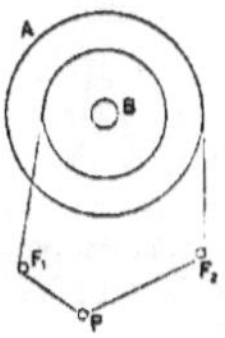

210. The following method of drawing a Cartesian by continuous motion was devised by Prof. Hammond:— A string is wound, as shown, around two pulleys turning on a common axis; a pencil at P holds the string taut around smooth pegs placed at random at F_1 and F_2; if the wheels be turned with the same angular velocity and the pencil does not slip on the string it will trace a Cartesian having F_1 and F_2 as foci.¹

Fig. 119.

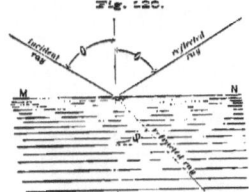

If the pulleys are *equal* the Cartesian will become an *ellipse*; if both threads are wound *the same way* around *either one* of the wheels the resulting curve will be an *hyperbola*.

211. It is a well-known fact that the incident and reflected ray make equal angles with the normal to a reflecting surface. If the latter is *curved* then each reflected ray cuts the one next to it, their consecutive intersections giving a curve called a *caustic by reflection*. Probably all have occasionally noticed such a curve on the surface of the milk in a glass, when the light was properly placed. If the reflecting curve is a *circle* the caustic is the *evolute of a limaçon*.

Fig. 120.

In passing from one medium *into* another, as from air into water, the deflection which a ray of light undergoes is called *refraction*, and for the same media the ratio of the *sines* of the angles of incidence and refraction (θ and ϕ, Fig. 120,) is constant. The consecutive intersections of *refracted* rays give also a *caustic*, which, for a *circle*, is the *evolute* of a *Cartesian Oval*. The proof of this statement² involves the property upon which is based the most convenient method of drawing a tangent to the Cartesian, viz., that the normal at any point divides the angle between the focal radii into parts whose sines are proportional to the factors of those radii in the equation. If, then, we have obtained a point G on the outer oval from the relation $m\rho' \pm n\rho'' = kc$ we may obtain the tangent at G by laying off on ρ' and ρ'' distances proportional to m and n, as $G r$ and $G h$, Fig. 118, then bisecting $r h$ at j and drawing the normal $G j$, to which the desired tangent is a perpendicular.

At a point on the inner oval the distance would not be laid off on a focal radius *produced*, as in the case illustrated.

¹ *American Journal of Mathematics*, 1878.   ² Salmon, *Higher Plane Curves*. Art. 117.

CASSIAN OVALS.

212. In the *Cassian Ovals* or *Ovals of Cassini* the points are connected with two foci by the relation $p' p'' = k^2$, i. e., the *product* of the focal radii is equal to some perfect square. These curves have already been alluded to in Art. 114 as plane sections of the annular torus, taken parallel to its axis.

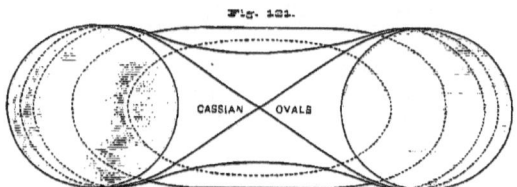

Fig. 121.

In Art. 158 one form—the Lemniscate—receives special treatment. For it the constant k^2 must equal m^2, the square of half the distance between the foci. When k is less than m the curve becomes two separate ovals.

213. The general construction depends on the fact that in any semicircle the square of an ordinate equals the product of the segments into which it divides the diameter. In Fig. 122 take F_1 and F_2 as the foci, erect a perpendicular $F_1 S$ to the axis $F_1 F_2$ and on it lay off $F_1 R$ equal to the constant k. Bisect $F_1 F_2$ at O and draw a semicircle of radius $O R$. This cuts the axis at A and B, the extreme points of the curve; for $k^2 = F_1 A \times F_1 B$. Any other point T may be obtained by drawing from F_1 a circular arc of radius $F_1 t$ greater than $F_1 A$; draw $t R$, then $R x$ perpendicular to it: $x F_1$ will then be the p'' and $F_1 t$ the p' for four points of the curve, which will be at the intersection of arcs struck from F_1 and F_2 as centres and with those radii.

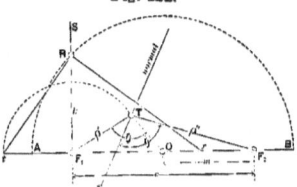

Fig. 122.

To get a normal at any point T draw $O T$, then make angle $F_2 T x = \theta = F_1 T O$; $T x$ will be the desired line.

THE CATENARY.

214. If a flexible chain, cable or string, of uniform weight per unit of length, be freely suspended by its extremities, the curve which it takes under the action of gravity is called a *Catenary*, from *catena, a chain*.

A simple and practical method of obtaining a catenary on the drawing-board would be to insert two pins in the board, in the desired relative position of the points of suspension, and then attach to them a string of the desired length. By holding the board vertically the string would assume the catenary, whose points could then be located with the pencil and joined in the usual manner with the irregular curve. Otherwise, if its points are to be located by means of an equation, we take axes in the plane of the curve, the y-axis (Fig. 123) being a vertical line through the lowest point T of the catenary, while the x-axis is a horizontal line at the distance m below T. The quantity m is called the *parameter* of the curve and is equal to the length of string which represents the tension at the lowest point.

THE CATENARY.—THE LOGARITHMIC SPIRAL.

The equation of the catenary[1] is then $y = \frac{m}{2}\left(e^{\frac{x}{m}} + e^{-\frac{x}{m}}\right)$ in which e is the *base* of Napierian logarithms[2] and has the numerical value 2.7182818 +.

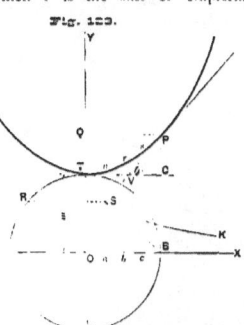

Fig. 123.

By taking successive values of x equal to m, $2m$, $3m$, etc., we get the following values for y:—

$x = m \ldots y = \frac{m}{2}\left(e + \frac{1}{e}\right)$ which for $m =$ *unity* becomes 1.54308

$x = 2m \ldots y = \frac{m}{2}\left(e^2 + \frac{1}{e^2}\right)$ " " 3.76217

$x = 3m \ldots y = \frac{m}{2}\left(e^3 + \frac{1}{e^3}\right)$ " " 10.0676

$x = 4m \ldots y = \frac{m}{2}\left(e^4 + \frac{1}{e^4}\right)$ " " 27.308

To construct the curve we therefore draw an arc of radius $OB = m$, giving T on the axis of y as the lowest point of the curve.

For $x = OB = m$ we have $y = BP = 1.54308$: for $x = Oa = \frac{m}{4}$ we have $y = an = 1.03142$.

The tension at any point P is equal to the weight of a piece of rope of length $BP = PC + m$. At the lowest point the tangent is horizontal. The length of any arc TP is proportional to the angle θ between TC and the tangent PV at the upper extremity of the arc.

215. If a circle RLB be drawn, of radius equal to m, it may be shown analytically that tangents PS and QR, to catenary and circle respectively, from points at the same level, will be parallel; also that PS equals the catenary-arc PrT; S therefore traces the involute of the catenary, and as SB always equals RO and remains perpendicular to PS (angle ORQ being always 90°) we have the curve TSK fulfilling the conditions of a *tractrix*. (See Art. 202.)

If a parabola, having a focal distance m, roll on a straight line, the focus will trace a catenary having m for its parameter.

The catenary was mistaken by Galileo for a parabola. In 1669 Jungius proved it to be neither a parabola nor hyperbola, but it was not till 1691 that its exact mathematical nature was known, being then established by James Bernouilli.

THE LOGARITHMIC OR EQUIANGULAR SPIRAL.

216. In Fig. 124 we have the curve called the *Logarithmic Spiral*. Its usual construction is based on the property that any radius vector, as ρ, which bisects the angle between two other radii, OM and ON, is a mean proportional between them; i. e. $\rho^2 = OS^2 = OM \times ON$.

If M and G are points of the spiral we may find an intermediate point K by drawing the ordinate OK to a semicircle of diameter $OM + OG$. A perpendicular through G to GK will then give D, another point of the curve, and this construction may be repeated indefinitely.

Radii making equal angles with each other are evidently in geometrical progression.

The curve never reaches the pole.

---

[1] Rankine. *Applied Mechanics*, Art. 175.
[2] In the expression $10^x = 100$ the quantity "2" is called the *logarithm* of 100, it being the exponent of the power to which 10 must be raised to give 100. Similarly 2 would be the logarithm of 64 were 8 the *base* or number to be raised to the power indicated.

76   THEORETICAL AND PRACTICAL GRAPHICS.

This spiral is often called *Equiangular* from the fact that the angle is always the same between a radius vector and the tangent at its extremity. Upon this property is based its use as the outline for spiral cams and for lobed wheels.

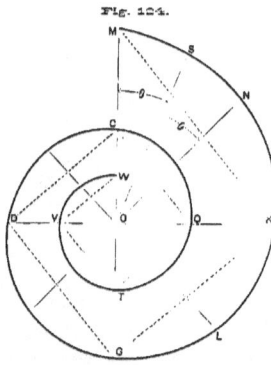

Fig. 124.

The name *logarithmic spiral* is based on the property that the angle of revolution is proportional to the logarithm of the radius vector. This is expressed by $\rho = a^\theta$, in which θ is the varying angle and a is some arbitrary constant.

To construct a tangent by calculation divide the hyperbolic logarithm[1] of the ratio $OM : OK$ (which are any two radii whose values are known) by the angle between these radii, expressed in circular measure[2]; the quotient will be the tangent of the constant angle of obliquity of the spiral.

217. Among the more interesting properties of this curve are the following:—

Its involute is an equal logarithmic spiral.

Were a light placed at the pole, the caustic—whether by reflection or refraction—would be a logarithmic spiral.

The discovery of these properties of recurrence led James Bernouilli to direct that this spiral be engraved on his tomb, with the inscription— *Eadem Mutata Resurgo*, which, freely translated, is— *I shall arise the same, though changed.*

Kepler discovered that the orbits of the planets and comets were conic sections having a focus at the centre of the sun. Newton proved that they would have described logarithmic spirals as they travelled out into space had the attraction of gravitation been inversely as the *cube* instead of the *square* of the distance.

THE HYPERBOLIC OR RECIPROCAL SPIRAL.

218. In this spiral the length of a radius vector is in inverse ratio to the angle through which it turns.

Like the logarithmic spiral it has an infinite number of convolutions about the pole, which it never reaches.

The invention of this curve is attributed to James Bernouilli, who showed that Newton's conclusions as to the logarithmic spiral (see Art. 217) would also hold for the hyperbolic spiral, the initial velocity of projection determining which trajectory was described. Fig. 125.

To obtain points of the curve divide a circle $m \, \vartheta \, s$ (Fig. 125) into any number of equal parts, and on some initial radius Om lay off some unit, as an inch; on the second radius $O\,2$ take $\frac{Om}{2}$; on the third $\frac{Om}{3}$; etc. For one-half the angle θ the radius vector would evidently be $2\,On$, giving a point s outside the circle.

The equation to the curve is $\frac{1}{r} = a\theta$, in which r is the

---
[1] To get the hyperbolic logarithm of a number multiply its common logarithm by 2.3026.
[2] In circular measure $360° = 2\pi r$ which for $r = 1$ becomes 6.28318; $180° = 3.14159$; $90° = 1.5708$; $60° = 1.0472$; $45° = 0.7854$; $30° = 0.5236$; $1° = 0.0174533$.

THE HYPERBOLIC SPIRAL.—THE LITUUS.

radius vector, a some numerical constant, and θ is the angular rotation of r (in circular measure) estimated from some initial line.

The curve has an asymptote parallel to the initial line and at a distance from it equal to $\frac{1}{a}$ units.

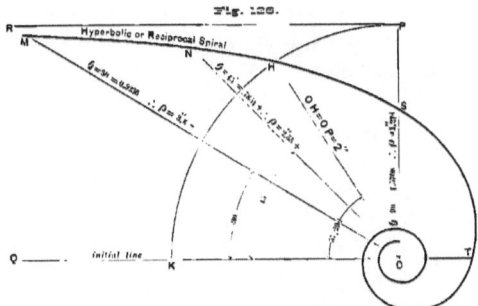

Fig. 126.

To construct the spiral from its equation take O as the pole (Fig. 126); OQ as the initial line; a for convenience, some fraction, as $\frac{1}{4}$; and as our *unit* some quantity, say half an inch, that will make $\frac{1}{a}$ of convenient size. Then taking QO as the initial line make $OP = \frac{1}{a} = 2''$ and draw PR parallel to OQ for the asymptote. For $\theta = 1$, that is, for *arc* KH = *radius* OH we have $r = \frac{1}{a} = 2''$, giving H, — one point of the spiral. Writing the equation in the form $r = \frac{1}{a} \cdot \frac{1}{\theta}$ and expressing various values of θ in circular measure we get the following:—

$\theta = 30° = 0.5236$; $r = OM = 3''.8 +$; $\theta = 45° = 0.7854$; $r = ON = 2''.55$;
$\theta = 90° = 1.5708$; $r = OS = 1''.2 +$; $\theta = 180° = 3.14159$; $r = OT = .6366$, etc.

The tangent to the curve at any point makes with the radius vector an angle ϕ which is found by analysis to sustain to the angle θ the following trigonometrical relation, $\tan \phi = \theta$; the circular measure of θ may therefore be found in a table of natural tangents and the corresponding value of ϕ obtained.

THE LITUUS.

219. The spiral in which the radius vector is inversely proportional to the square root of the angle through which it has revolved is called the *Lituus*. This relation is shown by the equation $r = \frac{1}{a\sqrt{\theta}}$, also written $a^2 \theta = \frac{1}{r^2}$.

When $\theta = 0$ we find $r = \infty$, making the initial line an asymptote to the curve.

In Fig. 127 take OQ as the initial line. O as the pole. $a = 2$, and our unit a three-inch line; then $\frac{1}{a} = 1\frac{1}{2}''$.

78   *THEORETICAL AND PRACTICAL GRAPHICS.*

For $\theta = 90° = \frac{\pi}{2}$ (in circular measure 1.5708) we have $r = OM = 1\overset{\prime\prime}{.}2 +$. For $\theta = 1$ we have the radius OT making an angle of $57°.29 +$ with the initial line, and in length equal to $\frac{1}{a}$ units, i. e., $1\frac{1}{4}''$. For $\theta = 45° = \frac{\pi}{4}$ (or 0.7854) r will be $OR = 1\overset{\prime\prime}{.}7 +$. Then $OH = \frac{OR}{2}$; for in rotating to OH the radius vector passes over four 45° angles, and the radius must therefore be one-half what it was for the first 45° described. Similarly $OK = \frac{OM}{2}$; $OM = \frac{OV}{2}$, etc.; this relation enabling the student to locate any number of points.

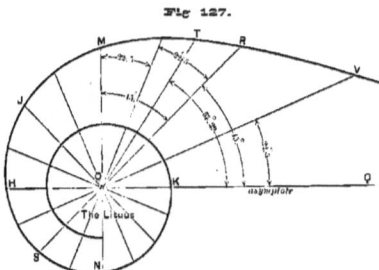

Fig. 127.

To draw a tangent to the curve we employ the relation $\tan \phi = 2\theta$, ϕ being the angle made by the tangent line with the radius vector, while θ is the angular rotation of the latter, in circular measure.

CHAPTER VI.

TINTING — FLAT AND GRADUATED. — MASONRY, TILING, WOOD GRAINING, RIVER-BEDS AND OTHER SECTIONS, WITH BRUSH ALONE OR IN COMBINED BRUSH AND LINE WORK.

220. Brush-work, with ink or colors, is either *flat* or *graduated*. The former gives the effect of a flat surface parallel to the paper on which the drawing is made, while graded tints either show curvature, or — if indicating flat surfaces — represent them as inclined to the paper, i. e., to the plane of projection. For either, the paper should be, as previously stated (Arts. 41 and 44) *cold-pressed* and *stretched*.

The surface to be tinted should not be abraded by sponge, knife or rubber.

221. The liquid employed for tinting must be free from sediment; or at least the latter, if present, must be allowed to settle, and the brush dipped only in the clear portion at the top. Tints may, therefore, best be mixed in an artist's water-glass, rather than in anything shallower. In case of several colors mixed together, however, it would be necessary to thoroughly stir up the tint each time before taking a brushful.

A tint prepared from a *cake* of high-grade India ink is far superior to any that can be made by using the ready-made liquid drawing inks.

222. The size of brush should bear some relation to that of the surface to be tinted; large brushes for large surfaces and vice versa. The customary error of beginners is to use too small and too dry a brush for tinting, and the reverse for shading.

223. Harsh outlines are to be avoided in brush work, especially in handsomely shaded drawings, in which, if sharply defined, they would detract from the general effect. This will become evident on comparing the spheres in Figs. 1 and 4 of Plate II.

Since tinting and shading can be successfully done, after a little practice, with only *pencilled* limits, there is but little excuse for inking the boundaries; but if, for the sake of definiteness, the outlines are inked at all it should be *before* the tinting, and in the finest of lines, preferably of "water-proof" ink; although any ink will do provided a soft sponge and plenty of clean water be applied to remove any excess that will "run." The sponge is also to be the main reliance of the draughtsman for the correction of errors in brush work; the water, however, and not the friction to be the active agent. An entire tint may be removed in this way in case it seems desirable.

224. When beginning work incline the board at a small angle, so that the tint will flow down after the brush. For a *flat*, that is, a *uniform* tint, start at the upper outline of the surface to be covered, and with a brush full, yet not surcharged — which would prevent its coming to a good point — pass lightly along from left to right, and on the return carry the tint down a little farther, making short, quick strokes, with the brush held almost perpendicular to the paper. Advance the tint as evenly as possible along a horizontal line; work quickly *between* outlines, but more slowly *along* outlines, as one should never overrun the latter and then resort to "trimming" to conceal lack of skill. It is possible for any one, with care and practice, to tint *to* yet not *over* boundaries.

The advancing edge of the tint must not be allowed to dry until the lower boundary is reached.

80   *THEORETICAL AND PRACTICAL GRAPHICS.*

No portion of the paper, however small, should be missed as the tint advances, as the work is likely to be spoiled by retouching.

Should any excess of tint be found along the lower edge of the figure it should be absorbed by the brush, after first removing the latter's surplus by means of blotting paper.

To get a dark effect several medium tints laid on in succession, each one drying before the next is applied, give better results than one dark one.

The heightened effect described in Art. 72, viz., a line of light on the upper and left-hand edges, may be obtained either (a) by ruling a broad line of *tint* with the drawing-pen at the desired distance from the outline, and instantly, before it dries, tinting from it with the brush; or (b) by ruling the line with the pen and thick Chinese White.

225. A tint will spread much more evenly on a large surface if the paper be first slightly dampened with clean water. As the tint will follow the water, the latter should be limited exactly to the intended outlines of the final tint.

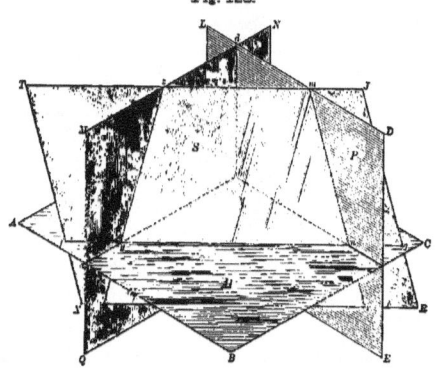
Fig. 128.

226. Of the colors frequently used by engineers and architects those which work best for flat effects are carmine, Prussian blue, burnt sienna and Payne's gray. Sepia and Gamboge, are, fortunately, rarely required for uniform tints; but the former works ideally for *shading* by the "dry" process described in the next article; and its rich brown gives effects unapproachable with anything else. It has, however, this peculiarity, that repeated touches upon a spot to make it darker produce the opposite effect, unless enough time elapses between the strokes to allow each addition to dry thoroughly.

227. For elementary practice with the brush the student should lay flat washes, in India tints, on from six to ten rectangles, of sizes between $2'' \times 6''$ and $6'' \times 10''$. If successful with these his next work may be the reproduction of Fig. 128, in which H, V, P and S denote horizontal, vertical, profile and section planes respectively. The figure should be considerably enlarged.

The plane V may have two washes of India ink; H one of Prussian blue; P one of burnt sienna, and S one of carmine.

The edges of the planes H, V and P are either vertical or inclined 30° to the horizontal.

For the section-plane assume n and m at pleasure, giving direction nm, to which JR and TX are parallel. A horizontal, mz, through m gives z. From n a horizontal, ny, gives y on ab. Joining y with z gives the "trace" of S on V.

228. Figures 129 and 130 illustrate the use of the brush in the representation of masonry. The former may be altogether in ink tints, or in medium burnt umber for the front rectangle of

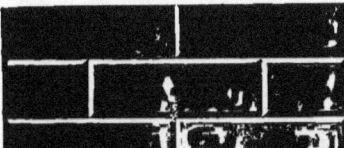

Fig. 129.

each stone, and dark tint of the same, directly from the cake, for the bevel. Lightly pencilled limits of bevel and rectangle will be needed; no inked outlines required or desirable.

The last remark applies also to Fig. 130, in which "quarry-faced" ashlar masonry is represented. If properly done, in either burnt umber or sepia, this gives a result of great beauty, especially effective on the piers of a large drawing of a bridge.

The darker portions are tinted directly from the cake, and are purposely made irregular and "jagged" to reproduce as closely as possible the fractured appearance of the stone.

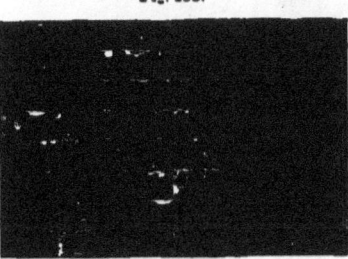

Fig. 130.

Two brushes are required when an "over-hang" or jutting portion is to be represented, one with a medium tint, the other with the thick color, as before. An irregular line being made with the latter, the tint is then softened out *on the lower side* with the point of the brush having the lighter tint. A light wash of the intended tone of the whole mass is quickly laid over each stone, either *before* or *after* the irregularities are represented, according as an exceedingly angular or a somewhat softened and rounded effect is desired.

82   THEORETICAL AND PRACTICAL GRAPHICS.

229. Designs in tiling are excellent exercises, not only for brush work in flat tints, but also—in their preliminary construction—in precision of line work. The superbly illustrated catalogues of the Minton Tile Works are, unfortunately, not accessible by all students, illustrating as they do, the finest and most varied work in this line, both of designer and chromo-lithographer; but it is quite within the bounds of possibility for the careful draughtsman to closely approach if not equal the standard and general appearance of their work, and as suggestions therefor Figs. 131 and 132 are presented.

230. In Fig. 131 the upper boundary, $a\,d\,h\,k$, of a rectangle is divided at a, b, c, etc., into equal spaces, and through each point of division two lines are drawn with the 30° triangle, as $b\,x$ and $b\,r$ through b. The oblique lines terminate on the sides and lower line of the rectangle. If the work is accurate—and it is worthless if not—any vertical line as $m\,n$, drawn through the intersection, m, of a pair of oblique lines, will pass through the intersection of a series of such pairs.

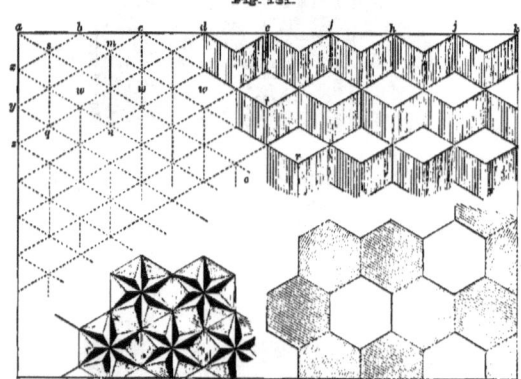

Fig. 131.

The figure shows three of the possible designs whose construction is based on the dotted lines of the figure. For that at the top and right, in which *horizontal* rows of rhombi are left white, we draw vertical lines as $s\,q$ and $m\,n$ from the lower vertex of each intended white rhombus, continuing it over two rhombi, when another white one will be reached. The dark faces of the design are to be finally in solid black, previous to which the lighter faces should be tinted with some drab or brown tint. The pencilled construction lines would necessarily be erased before the tint was laid on.

The most opaque effect in colors is obtained by mixing a large portion of Chinese white with the water color, making what is called by artists a "body color." Such a mixture gives a result in marked contrast with the transparent effect of the usual wash; but the amount of white used should be sufficient to make the tint in reality a *paste*, and no more should be taken on the brush at one time than is needed to cover one figure.

Sepia and Chinese white, mixed in the proper proportions, give a tint which contrasts most agreeably with the black and white of the remainder of the figure. The star design and the hexagons in the lower right-hand corner result from extensions or modifications of the construction just described which will become evident on careful inspection.

TINTING.—BRUSH SHADING.

231. Fig. 132 is a Minton design with which many are familiar, and which affords opportunity for considerable variety in finish. Its construction is almost self-evident. The *equal* spaces, $a\,b$, $c\,d$, $m\,n$—which may be any width, x,—alternate with other equal spaces $b\,c$, which may preferably be about $3\,x$ in width. Lines at $45°$, as indicated, complete the preliminaries to tinting.

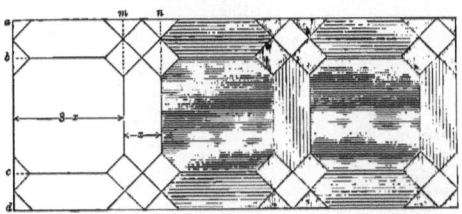

Fig. 132.

The octagons may be in Prussian blue, the hexagons in carmine, and the remainder in white and black, as shown; or browns and drabs may be employed for more subdued effects.

SHADING.

232. For *shading*, by graduated tints, provide a glass of clear water in addition to the tint; also an ample supply of blotting paper.

The water-color or ink tint may be considerably darker than for flat tinting; in fact, the darker it is, provided it is clear, the more rapidly can the desired effect be obtained.

The brush must contain much less liquid than for flat work.

Lay a narrow band of tint quickly along the part that is to be the darkest, then dip the brush into clear water and immediately apply it to the blotter, both to bring it to a good point and to remove the surplus tint. With the now once-diluted tint carry the advancing edge of the band slightly farther. Repeat the operation until the tint is no longer discernible as such.

The process may be repeated from the same starting point as many times as necessary to produce the desired effect; but the work should be allowed to dry each time before laying on a new tint.

Any irregularities or streaks can easily be removed after the work dries, by retouching or "stippling" with the point of a fine brush that contains *but little tint*—scarcely more than enough to enable the brush to retain its point. For small work, as the shading of rivets, rods, etc., the process just mentioned, which is also called "dry shading," is especially adapted, and, although somewhat tedious, gives the handsomest effects possible to the draughtsman.

233. Where a good, *general* effect is wanted, to be obtained in less time than would be required for the preceding processes, the method of over-lapping flat tints may be adopted. A narrower band of dark tint is first laid over the part to be the darkest. When dry this is overlaid by a broader band of lighter tint. A yet lighter wash follows, beginning on the dark portion and extending still farther than its predecessor. The process is repeated with further diluted tints until the desired effect is obtained.

Faintly-penciled lines may be drawn at the outset as limits for the edges of the tints.

84   *THEORETICAL AND PRACTICAL GRAPHICS.*

This method is better adapted for large work, that is not to be closely scrutinized, than for drawings that deserve a high degree of finish.

234. As to the relative position and gradation of the lights and shades on a figure, the student is referred to Arts. 78 and 79 and the chapter on shadows; also to the figures of Plate II, which may serve as examples to be imitated while the learner is acquiring facility in the use of the brush, and before entering upon constructive work in shades and shadows. Fig. 3 of Plate II may be undertaken first, and the contrast made yet greater between the upper and lower boundaries. Fig. 1 (Plate II) requires no explanation. In Fig. 133 we have a wood-cut of a sphere, with the theoretical dark or "shade" line more sharply defined than in the spheres on the plate.

Fig. 133.

Fig. 134.

A drawing of the end of a highly-polished revolving shaft, or even of an ordinary metallic disc, would be shaded as in Fig. 134.

Fig. 2 (Plate II) represents the triangular-threaded screw, its oblique surfaces being, in mathematical language, *warped helicoids*, generated by a moving straight line, one end of which travels *along the axis* of a cylinder while the other end traces or follows a *helix* on the cylinder.

The construction of the helix having already been given (Art. 120) the outlines can readily be drawn. The method of exactly locating the shadow and shade lines will be found in the chapter on shadows.

Fig. 4 (Plate II), when compared with Fig. 91, illustrates the possibilities as to the representation of interesting mathematical relations. The fact may again be mentioned, on the principle of "line upon line," as also for the benefit of any who may not have read all that has preceded, that the spheres in the cone are tangent to the oblique plane at the *foci* of the elliptical section. The peculiar dotted effect in this figure is due to the fact that the original drawing, of which this is a photographic reproduction by the gelatine process, was made with a lithographic crayon upon a special pebbled paper much used by lithographers. The original of Fig. 1, on the other hand, was a brush-shaded sphere on Whatman's paper.

235. Fig. 5 (Plate II) shows a "Phœnix column," the strongest form of iron for a given weight,

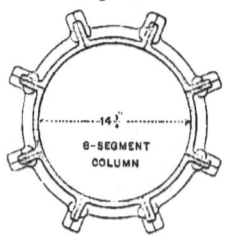
Fig. 135.

for sustaining compression. The student is familiar with it as an element of outdoor construction in bridges, elevated railroads, etc.; also in indoor work in many of the higher office buildings of our great cities.

By drawing first an end view of a Phœnix column, similar to that of Fig. 135, we can readily derive an oblique view like that of the plate, by including it between parallels from all points of the former. The proportions of the columns are obtainable from the tables of the company.

Fig. 135 is a cross-section of the 8-segment column, the shaded portion showing the minimum and the other lines the maximum size for the same inside diameter.

MATERIALS OF CONSTRUCTION.

In a later chapter the proportions of other forms of structural iron will be found. Short lengths of any of these, if shown in oblique view, are good subjects for the brush, especially for "dry" shading, the effect to be aimed at being that of the rail section of Fig. 136.

Fig. 136.

236. When some particular material is to be indicated, a flat tint of the proper technical color (see Art. 73) should be laid on with the brush, either before or after shading. When the latter is done with sepia it is probably safer to lay on the flat tint first.

A darker tint of the technical color should always be given to a cross-section. For blue-printing, a cross-section may be indicated in solid black.

WOOD.—RIVER-BEDS.—MASONRY, ETC.

237. While the engineering draughtsman is ordinarily so pressed for time as not to be able to give his work the highest finish, yet he ought to be able, when occasion demands, to obtain both natural and artistic effects; and to conduce to that end the writer has taken pains to illustrate a number of ways of representing the materials of construction. Although nearly all of them may be—and in the cuts are—represented in black and white (with the exception of the wood-graining on Plate II), yet colors, in combined brush and line work, are preferable. The student will, however, need considerable practice with pen and ink before it will be worth while to work on a tinted figure.

238. Ordinarily, in representing wood, the mere fact that it *is* wood is all that is intended to be indicated. This may be done most simply by a series of irregular, approximately-parallel lines, as in Fig. 10 or as on the rule in Fig. 17, page 12. Make no attempt, however, to have the grain *very* irregular. The natural unsteadiness of the hand, in drawing a long line toward one continuously, will cause almost all the irregularity desired.

If a better effect is wanted, yet without color, the lines may be as in Fig. 107, which represents hard wood.

In graining, the draughtsman should make his lines *toward* himself, standing, so to speak, at the end of the plank upon which he is working.

The splintered end of a plank should be sharply toothed, in contradistinction to a metal or stone fracture, which is what might be called smoothly irregular.

239. An examination of any piece of wood on which the grain is at all marked will show that it is darker at the inner vertex of any marking than at the outer point. Although this difference is more readily produced with the brush, yet it may be shown in a satisfactory degree with the pen, by a series of after-touches.

240. If we fill the pen with a rather dark tint of the conventional color, draw the grain as in the figures just referred to, and then overlay all with a medium flat wash of some properly chosen color, we get effects similar to those of Plate II.

On large timber-work the preliminary graining, as also the final wash, may be done altogether with the brush; as was the original of Fig. 9, Plate II.

End views of timbers and planks are conventionally represented by a series of concentric free-hand rings in which the spacing increases with the distance from the heart; these are overlaid with a few radial strokes of darker tint. In ink alone the appearance is shown in Figs. 39 and 115.

241. The color-mixtures recommended by different writers on wood graining are something short of infinite in number; but with the addition of one or two colors to those listed in the draughtsman's outfit (Art. 56) one should be able to imitate nature's tints very closely.

86   THEORETICAL AND PRACTICAL GRAPHICS.

No hard-and-fast rule as to the proportions of the colors can be given. In this connection we may quote Sir Joshua Reynolds' reply to the one who inquired how he mixed his paints. "With brains," said he. One *general* rule, however; always employ delicate rather than glaring tints.

Merely to indicate wood with a color and no graining use burnt sienna, the tint of Figs 7, 8 and 10 of Plate II.

Drawing from the writer's experience and from the suggestions of various experimenters in this line the following hints are presented:—

In every case grain first, then overlay with the ground tint, which should always be much lighter than the color used for the grain. If possible have at hand a good specimen of the wood to be imitated.

Hard Pine: Grain—burnt umber with either carmine or crimson lake; for overlay add a little gamboge to the grain-tint diluted.

Soft Pine: Gamboge or yellow ochre with a small amount of burnt sienna.

Black Walnut: Grain—burnt umber and a very little dragon's blood; final overlay of modified tint of the same or with the addition of Payne's gray.

Fig. 137.

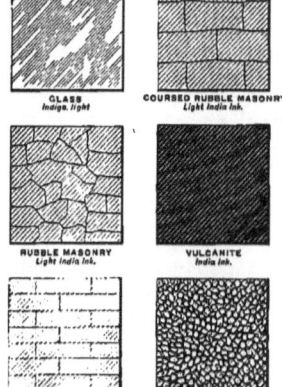

GLASS
Indigo, light

COURSED RUBBLE MASONRY
Light India Ink.

RUBBLE MASONRY
Light India Ink.

VULCANITE
India Ink.

BRICK
Venetian Red.

CONCRETE
Yellow Ochre.

Oak: Grain—burnt sienna; for overlay, the same, with yellow ochre.

Chestnut: Grain—burnt umber and dragon's blood; overlay of the same, diluted, and with a large proportion of gamboge or light yellow added.

Spruce: Grain—burnt umber, medium; add yellow ochre for the overlay.

Mahogany: Grain—burnt sienna or umber with a small amount of dragon's blood; dilute, and add light yellow for the overlay.

Rosewood: Grain—replace the dragon's blood of mahogany-grain by carmine, and for overlay dilute and add a little Prussian blue.

242. *River-beds* in black and white or in colors have been already treated in Art. 26, to which it is only necessary to add that such sections are usually made quite narrow, and, preferably—if in color—shaded quite abruptly on the side opposite the water.

243. The sections of *masonry*, *concrete*, *brick*, *glass* and *vulcanite*, given on page 25 as pen and ink exercises, are again

Fig. 138.

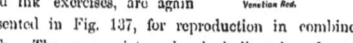

presented in Fig. 137, for reproduction in combined brush and line work. The appropriate color is indicated under each section.

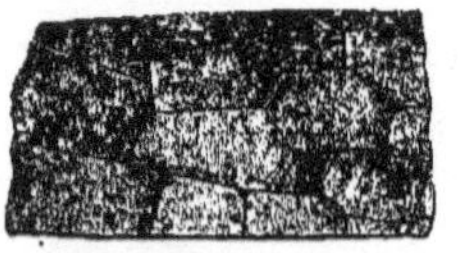

244. *Masonry* constructions may be broadly divided into *rubble* and *ashlar*.

In *ashlar* masonry the bed-surfaces and the joints (edges) are shaped and dressed with great care, so that the stones may not only be placed in regular layers or *courses*, but often fill exactly some predetermined place, as in arch construction, in which case the determination of their forms and the derivation of the patterns for the stone-cutter involves the application of the Descriptive Geometry of Monge. (Art. 283).

REPRESENTATION OF MASONRY.

Rubble work, however, consists of constructions involving stones mainly "in the rough," but may be either coursed or uncoursed. Fig. 138 is a neat example of uncoursed though partially dressed or "hammered" rubble. In section, as shown in Fig. 137, it is merely necessary to rule section-lines over the boundaries of the stones—a remark applying equally to ashlar masonry.

Fig. 139.

The other examples in this chapter are of ashlar, mainly "quarry-faced," that is, with the front nearly as rough as when quarried. A beveled or "chamfered" ashlar is shown in Figs. 129 and 140, the latter shaded in what is probably the most effective way for small work, viz., with *dots*, the effect depending upon the *number*, not the *size* of the latter.

Fig. 140.

Only a careful examination of the *kind* and *position* of the lines in the other figures on this page will disclose the secret of the variety in the effects produced. For the handsomest results with any of these figures the pen-work—

Fig. 141.

Fig. 142.

whether dotting or "cross-hatching"—should be preceded by an undertone of either India ink, umber, Payne's gray, cobalt or Prussian blue, according to the kind of stone to be represented.

Fig. 143.

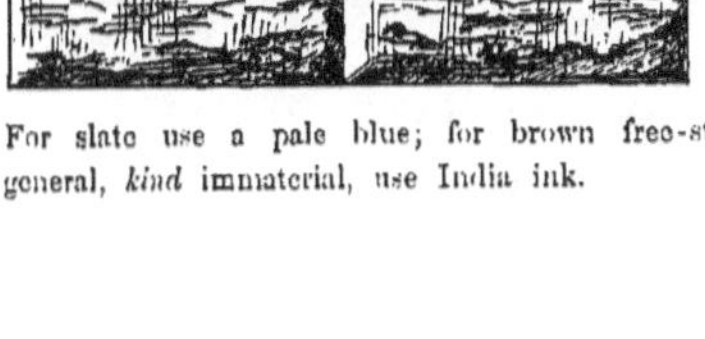

Fig. 144.

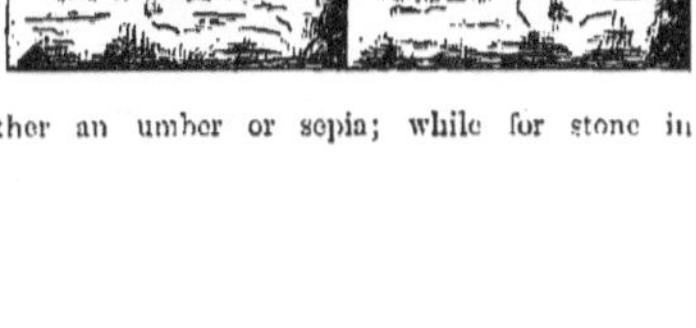

For slate use a pale blue; for brown free-stone either an umber or sepia; while for stone in general, *kind* immaterial, use India ink.

CHAPTER VII.

FREE-HAND AND MECHANICAL LETTERING.—PROPORTIONING OF TITLES.

245. Practice in lettering forms an essential part of the elementary work of a draughtsman. Every drawing has to have its title, and the general effect of the result as a whole depends largely upon the quality of the lettering.

Other things being equal, the expert and rapid draughtsman in this line has a great advantage over one who can do it but slowly. For this reason free-hand lettering is at a high premium, and the beginner should, therefore, aim not only to have his letters *correctly formed* and *properly spaced*, but, as far as possible, to do without mechanical aids in their construction. When under great pressure as to time it is, however, perfectly legitimate to employ some of the mechanical expedients used in large establishments as "short cuts" and labor-savers. Among these the principal are "tracing" and the use of rubber types.

246. To *trace* a title one must have at hand complete printed alphabets *of the size of type required*. Placing a piece of tracing-paper over the letter wanted, it is traced with a hard pencil, the paper then slipped along to the next letter needed, and the process repeated until the words desired have been outlined. The title is then transferred to the drawing by first running over the lines on the back of the tracing-paper with a soft pencil, after which it is only necessary to re-trace the letters with a hard pencil, on the face of the transfer-paper, to find their outlines faintly yet sufficiently indicated on the paper underneath. Carbon paper may also be used for transferring.

247. The process just described would be of little service to a ready free-hand draughtsman, but with the use of *rubber types*, for the words most frequently recurring in the titles, a merely average worker may easily get results which—in point of time—cannot be exceeded by any other method. When employing such types either of the following ways may be adopted: (a) a light impression may be made with the aniline ink ordinarily used on the pads, and the outlines then followed and the "filling in" done either with a writing-pen* or fine-pointed sable-hair brush; or (b) the impression may be made after moistening the types on a pad that has been thoroughly wet with a light tint of India ink. The drawing-ink must then be immediately applied, free-hand, with a Falcon pen or sable brush, before the type-impression can dry. The pen need only be passed down the middle of a line, as on the dampened surface the ink will spread instantly to the outlines.

248. The educated draughtsman should, however, be able not only to draw a legible title of the simple character required for shop-work, and in which the foregoing expedients would be mainly serviceable, but be prepared also for work out of the ordinary line, and, if need be, quite elaborate, as on a competitive drawing. Such knowledge can only be gained by careful observation of the forms of letters, and considerable practice in their construction.

No rigid rules can be laid down as to choice of alphabets for the various possible cases. Common-sense, custom and a natural regard for the "fitness of things" are the determining factors.

Obviously rustic letters would be out of place on a geometrical drawing, and other incongruities

---

* Refer to Art. 27 with regard to the pens to be used for the various styles of letters.

will naturally suggest themselves. In addition to the hints in Art. 27 a few general principles and methods may, however, be stated to the advantage of the beginner, who should also refer to the special instructions given in connection with certain specimen alphabets at the end of this work.

249. In the first place, a title should be symmetrical with respect to a vertical centre-line, a rule which should be violated but rarely, and then, usually, when the title is to be somewhat fancy in design, as for a magazine cover.

Elementary Plates
in
MECHANICAL DRAWING
drawn by Cortlandt Van Corlear at the
LEADING TECHNICAL SCHOOL
Jan.-June, 3001.

250. If it be a *complete* as distinguished from a *partial* or *sub*-title it will answer the following questions which would naturally arise in the mind of the examiner:—
What is it?—Where done?—By whom?—When?—On what scale?

In answering these questions the relative valuation and importance of the lines are expressed by the *sizes* and *kinds* of type chosen. This is a point requiring most careful consideration, as the final effect depends largely upon a proper balancing of values.

DETAIL DRAWINGS
OF
PERFECTION SUSPENSION BRIDGE
designed by
Goodwin, Mackenzie & Cartwright
MINNEAPOLIS, MINN.
SCALE 4 FT. / 1 IN. June 16, 2900. JOSE MARTINEZ, DEL.

251. The "By whom?" may cover two possibilities. In the case of a set of drawings made in a scientific school it would refer to the *draughtsman*, and his name might properly have considerably greater prominence than in any other case. The upper title on this page is illustrative of this point, as also of a symmetrical and balanced arrangement, although cramped as to space, vertically.

Ordinarily the "By whom?" will refer to the designer, and the draughtsman's name ought to be comparatively inconspicuous, while the name of the designer should be given a fair degree of prominence. This, and other important points to be mentioned, are illustrated in the preceding arrangement, printed, like the upper title, from types of which complete alphabets will be found at the end of this work.

252. The abbreviation *Del.*, often placed after the draughtsman's name, is for *Delineavit*—*He drew it*—and does not indicate what the visitor at the exhibition supposed, that all good draughtsmen hail from Delaware.

253. The best designed titles are either in the form of two truncated pyramids having, if possible, the most important line as their common base, or else elliptical in shape.

254. The use of capitals throughout a line depends upon the style of type. It gives a most unsatisfactory result if the letters are of irregular outline, as is amply evidenced by the words

𝔐𝔢𝔠𝔥𝔞𝔫𝔦𝔠𝔞𝔩 𝔇𝔯𝔞𝔴𝔦𝔫𝔤,

each letter of which is exquisite in form, but the combination almost illegible. Contrast them with the same style, but in capitals and small letters:—

𝔐𝔢𝔠𝔥𝔞𝔫𝔦𝔠𝔞𝔩 𝔇𝔯𝔞𝔴𝔦𝔫𝔤.

255. As to *spacing*, the visible white spaces between the letters should be as nearly the same as possible. In this feature, as in others, the draughtsman can get much more pleasing results than the printer, since the latter usually has each letter on a separate piece of metal, and can not adjust his space to any particular combination of letters, such as F A, L V, W A or A V, where a better effect would be obtained by placing the lower part of one letter under the upper part of the next. This is illustrated in Fig. 146, which may be contrasted with the printer's best spacing of the separate types for the A and W in the word "Drawings" of the last title.

Fig. 146.

256. The *amount of space* between letters will depend upon the length of line that the word or words must make. If an important word has few letters they should be "spaced out," and the letters themselves of the "extended" kind, i. e., broader than their height. The following word will illustrate. The characteristic feature of this type, viz., heavy horizontals and light verticals, is common to all the variations of a fundamental form frequently called *Italian Print*.

B R I D G E.

When, on the other hand, many letters must be crowded into a small space, a "condensed" style of letter must be adopted, of which the following is an example:

Pennsylvania Railroad.

257. While the varieties of letters are very numerous yet they are all but changes rung on a few fundamental or basal forms, the most elementary of which is the

GOTHIC, ALSO CALLED HALF-BLOCK.

Letters like B, O, etc., which have, usually, either few straight parts or none at all, may, for the sake of variety as also for convenience of construction, be made partially or wholly angular; in the latter case the form is called *Geometric Gothic* by some type manufacturers. It is only appropriate for work exclusively mechanical. The rounded forms are preferable for free-hand lettering.

LETTERING.

The following complete Gothic alphabet is so constructed that whether designed in its "condensed" or "extended" form the proper proportions may be easily preserved.

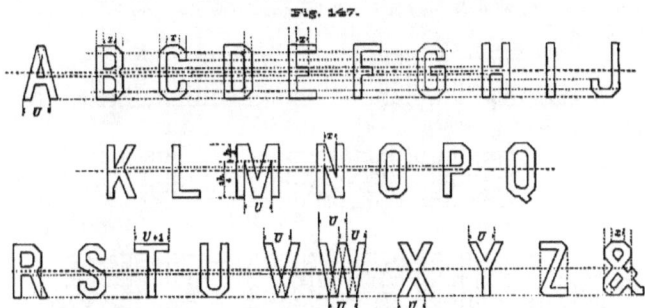

Fig. 147.

Taking all the solid parts of the letters at the same width as the I, we will find any letter *of average width*, as U, to be twice that unit, plus the opening between the uprights, which last, being indeterminate, we may call x, making it small for a "condensed" letter, and broad as need be for an "extended" form.

The word *march* would foot up $5U + 3$, disregarding—as we would invariably—the amount the foot of the R projects beyond the main right-hand outline of the letter. In terms of x this makes $5x + 13$, as $U = x + 2$. Allowing spaces of $1\frac{1}{4}$ unit width between letters adds 5 to the above, making $5x + 18$ for the total length in terms of the I. Assuming x equal to twice the unit we would have the whole word equal to twenty-eight units; and if it were to extend seven inches the width of the solid parts would therefore be one-quarter of an inch.

Where the width of a letter is not indicated it is assumed to be that of the U. The W is equal to $2U - 1$. This relation, however, does not hold good in all alphabets.

The angular corners are drawn usually with the 45° triangle.

The guide-lines show what points of the various letters are to be found on the same level, and should be but faintly pencilled.

As remarked in Art. 27, the extended form of Gothic is one of the best for dimensioning and lettering *working drawings*, and is rapidly coming into use by the profession.

258. The Full-Block letter next illustrated is easier to work with than the Gothic in the matter of preliminary estimate, as the width of each letter—in terms of unit squares—is evident at a glance.

The same word *march* would foot up twenty-seven squares without allowing for spaces between letters. Calling the latter each *two* we would have thirty-five squares for the same length as before (seven inches), making one-fifth of an inch for the width of the solid parts. For convenience the widths of the various letters are summarized:

I = 3; C, G, O, Q, S, Z = 4; A, B, D, E, F, J, L, P, R, T, & = 5; H, K, N, U, V, X, Y = 6; M = 7; W = 8.

259. In case the preliminary figuring were only approximate and there were but two words in the line, as, for example, *Mechanical Drawing*, a safe method of working would be to make a fair allowance for the space between the words, begin the first word at the calculated distance to the left of the vertical centre-line, complete it, then work the second word backward, beginning with the

92   *THEORETICAL AND PRACTICAL GRAPHICS.*

G as far to the right of the reference line as the M was to the left. On completing the second word any difference between the actual and the estimated length of the words, due to over- or under-width of such letters as M, W and I, will be merged into the space between the words.

Fig. 148.

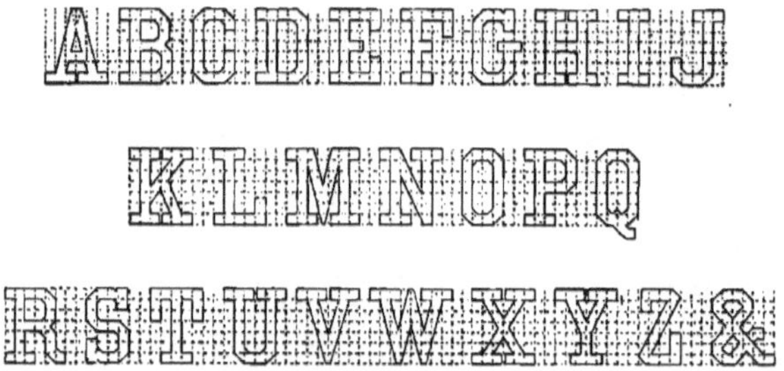

ABCDEFGHIJ
KLMNOPQ
RSTUVWXYZ&

With three words in a line the same method might be adopted, the middle word being easily placed half way between the others, which, by this method of construction would not only begin correctly but also terminate where they should.

260. Note particularly that the top of a B is always slightly smaller than the bottom; *Fig. 149.* similarly with the S. This is made necessary by the fact that the eye seems to exaggerate the upper half of a letter. To get an idea of the amount of difference allowable compare the following equal letters printed from Roman type, condensed. Although not so important in the E, some difference between top and bottom may still to advantage be made. Another refinement is the location of the horizontal cross-bar of an A slightly below the middle of the letter.

SS

261. While vertical letters are most frequently used, yet no handsomer effect can be obtained than by a well-executed inclined letter. The angle of inclination should be about 70°.

Beginners usually fail sadly in their first attempt with the A and V, one of whose sides they give the same slant as the upright of the other letters. In point of fact, however, it is the imaginary (though, in the construction, pencilled) *centre-line* which should have that inclination. See Fig. 150.

Fig. 150.

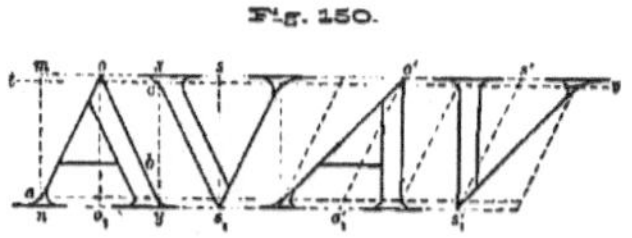

In these forms—the Roman and Italic Roman—the union of the light horizontals or "seriffs" with the other parts is in general effected by means of fine arcs, called "fillets," drawn free-hand. On many letters of this alphabet *some* lines will, however, meet at an angle, and only a careful examination of good models will enable one to construct correct forms. Upon the size of the fillets the appearance of the letter mainly depends, as will be seen by a glance at Fig. 151, which reproduces, exactly, the N of each of two leading alphabet books. If the fillets round out to the end of the spur of the letter, a coarse and bulky appearance is evidently the result; while a fine curve, leaving the straight horizontals projecting beyond them, gives the finish desired. This is further illustrated by No. 23 of the alphabets appended, a type which for clearness and elegance is a triumph of the founder's art. As usually constructed, however, the D and R are finished at the top like the P.

Fig. 151.

NN

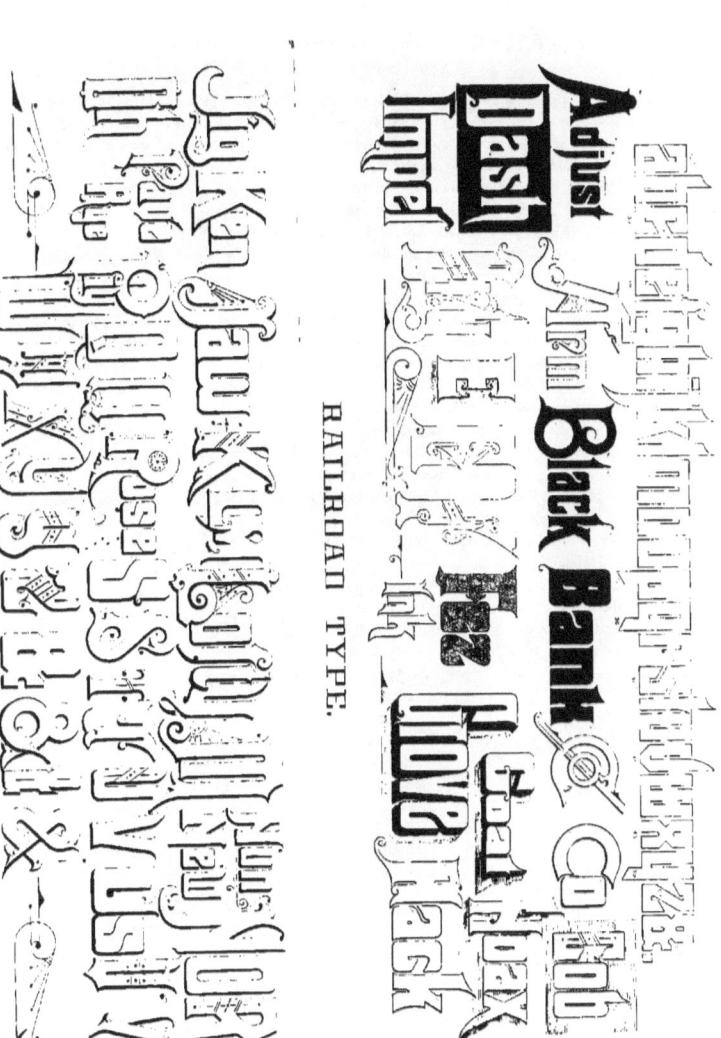

RAILROAD TYPE.

262. The Roman alphabet and its inclined or *italic* form are much used in topographical work.

A text-book devoted entirely to the Roman alphabet is in the market, and in some works on topographical drawing very elaborate tables of proportions for the letters are presented; these answer admirably for the construction of a standard alphabet, but in practice the proportions of the model would be preserved by the draughtsman no more closely than his eye could secure. Usually the small letters should be about three-fifths the height of the capitals. Except when more than one-third of an inch in height, these letters should be entirely free-hand.

263. *When a line of a title is curved* no change is made in the *forms* of the letters; but if of a *vertical*, as distinguished from a slanting or *italic* type, the centre-line of each letter should, if produced, pass through the centre of the curve.

Italic letters, when arranged on a curve, should have their centre-lines inclined at the same angle to the normal (or radius) of the curve as they ordinarily make with the vertical.

264. An alphabet which gives a most satisfactory appearance, yet can be constructed with great rapidity, is what we may call the "Railroad" type, since the public has become familiar with it mainly from its frequent use in railroad advertisements.

The fundamental forms of the small letters, with the essential construction lines, are given in rectangular outline in the complete alphabet on the preceding page, with various modifications thereof in the words below them, showing a large number of possible effects.

At least one plain and fancy capital of each letter is also to be found on the same page, with in some instances a still larger range of choice.

No handsomer effects are obtainable than with this alphabet, when brush tints are employed for the undertone and shadows.

265. For rapid lettering on tracing-cloth, Bristol board or any smooth-surfaced paper a style long used abroad and increasing in favor in this country is that known as *Round Writing*, illustrated by Fig. 152, and for which a special text-book and pens have been prepared by F. Soennecken. The pens are stubs of various widths, cut off obliquely, and when in use should not, as ordinarily, be dipped into the ink, but the latter should be inserted, by means of another pen, between the *top* of the Soennecken pen and the brass "feeder" that is usually slipped over it to regulate the flow.

The Soennecken Round Writing Pens are also by far the best for lettering in *Old English*, *German Text* and kindred types.

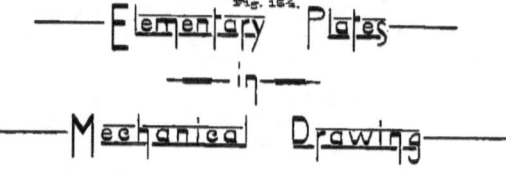

The improvement due to the addition of a few straight lines to an ordinary title will become evident by comparing Figs. 153 and 154. The judicious use of "word ornaments," such as those of alphabets 33, 42, 49, and of several of the other forms illustrated, will greatly enhance the appearance of a title without materially increasing the time expended on it. This is illustrated in the lower title on page 89.

DESIGNS FOR BORDERS.

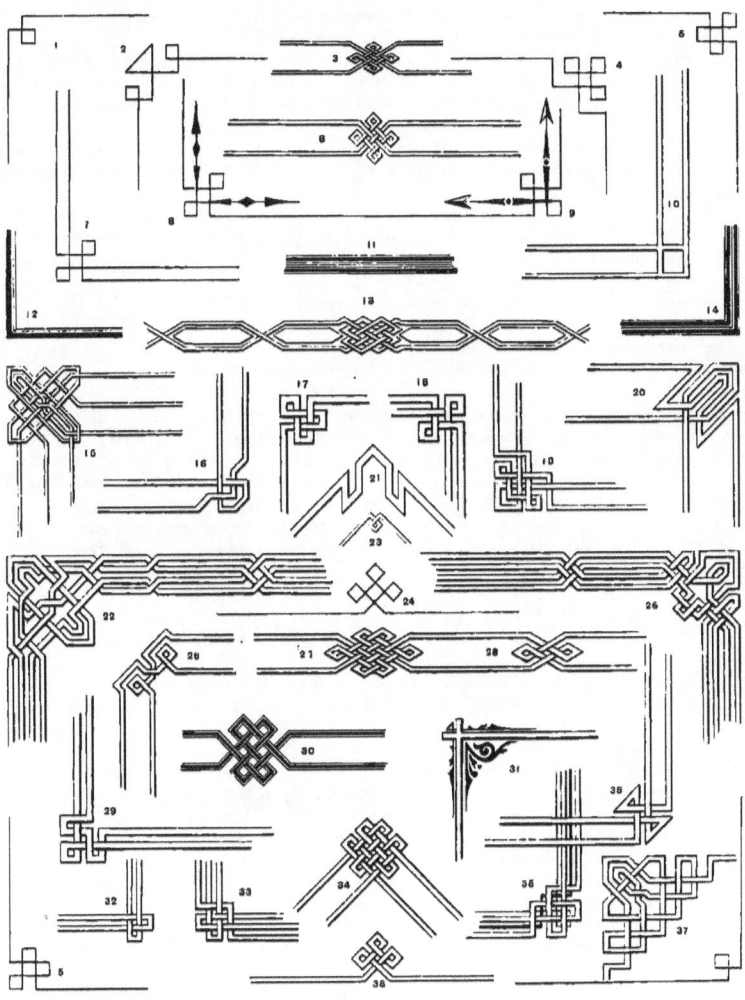

96   THEORETICAL AND PRACTICAL GRAPHICS.

266. *Borders.* Another effective adjunct to a map or other drawing is a neat border. It should be strictly in keeping with the drawing, both as to character and simplicity.

On page 95 a large number of corner designs and borders is presented, one-third of them original designs, by the writer, for this work. The principle of their construction is illustrated by Fig. 155, in which the larger design shows the necessary preliminary lines, and the smaller the complete corner. It is evident in this, as in all cases of interlaced designs, that we must first lay off each way from the corner as many equal distances as there are bands and spaces, and lightly make a network of squares — or of rhombi, if the angles are acute — by pencilled construction-lines through the points of division.

267. *Shade lines on borders.* The usual rule as to shade lines applies equally to these designs, thus: Following any band or pair of lines making the turns as one piece, if it runs *horizontally* the *lower* line is the heavier, while in a *vertical* pair the *right-hand* line is the shaded line. This is on the assumption that the light is coming in the direction usually assumed for mechanical drawings, i. e., descending diagonally from left to right.

In case a pair of lines runs obliquely, the shaded lines may be determined by a study of their location on the designs of the plate of borders.

It need hardly be said that on any drawing and its title the light should be supposed to come from *but one direction throughout*, and not be shifted; and the shaded lines should be located accordingly. This rule is always imperative.

In drawing for scientific illustration or in art work it is allowable to depart from the usual strictly conventional direction of light, if a better effect can thereby be secured.

268. A striking letter can be made by drawing the shade line only, as in Fig. 146, page 90, which we may call "Full-Block Shade-Line," being based upon the alphabet of Fig. 148, page 92, as to construction. Owing to its having more projecting parts it gives a much handsomer effect than the

HALF-BLOCK SHADE-LINE.

The student will notice that the light comes from different directions in the two examples.

These forms are to the ordinary fully-outlined letters what art work of the "impressionist" school is to the extremely detailed and painstaking work of many; what is actually seen suggests an equal amount not on the paper or canvas.

269. While a teacher of draughting may well have on hand, as reference works for his class, such books on lettering as Prang's, Becker's and others equally elaborate, yet they will be found of only occasional service, their designs being as a rule more highly ornate than any but the specialist would dare undertake, and mainly of a character unsuitable for the usual work of the engineering or architectural draughtsman, whose needs were especially in mind when selecting types for this work.

The alphabets appended afford a large range of choice among the handsomest forms recently designed by the leading type manufacturers, also containing the best among former types; and with the "Railroad," Full-Block and Half-Block alphabets of this chapter, proportioned and drawn by the writer, supply the student with a practical "stock in trade" that it is believed will require but little, if any, supplementing.

CHAPTER VIII.

BLUE-PRINT AND OTHER COPYING PROCESSES.—METHODS OF ILLUSTRATION.

270. While in a draughting office the process described below is, at present, the only method of copying drawings with which it is *absolutely essential* that the draughtsman should be thoroughly acquainted, he may, nevertheless, find it to his advantage to know how to prepare drawings for reproduction by some of the other methods in most general use. He ought also to be able to recognize, usually, by a glance at an illustration, the method by which it was obtained. Some brief hints on these points are therefore introduced.

This is, obviously, however, not the place to give full particulars as to all these processes, even were the methods of manipulation not, in some cases, still "trade secrets"; but all important details concerning them, that have become common property, may be obtained from the following valuable works: *Modern Heliographic Processes*,[*] by Ernst Lietze; *Photo-Engraving, Etching and Lithography*,[†] by W. T. Wilkinson; and *Modern Reproductive Graphic Processes*,[*] by Jas. S. Pettit.

THE BLUE-PRINT PROCESS.

271. By means of this process, invented by Sir John Herschel, any number of copies of a drawing can be made, in white lines on a blue ground. In Arts. 43 and 45 some hints will be found as to the relative merits of tracing-cloth and "Bond" paper, for the original drawing.

A sheet of paper may be sensitized to the action of light by coating its surface with a solution of red prussiate of potash (ferrocyanide of potassium) and a ferric salt. The chemical action of light upon this is the production of a ferrous salt from the ferric compound; this combines with the ferrocyanide to produce the final blue undertone of the sheet; while the portions of the paper, from which the light was intercepted by the lines of the drawing, become white after immersion in water.

The proportions in which the chemicals are to be mixed, are, apparently, a matter of indifference, so great is the disparity between the recipes of different writers; indeed, one successful draughtsman says: "Almost any proportion of chemicals will make blue-prints." Whichever recipe is adopted—and a considerable range of choice will be found in this chapter—the hints immediately following are of general application.

272. Any white paper will do for sensitizing that has a hard finish, like that of ledger paper, so as not to absorb the chemical solution.

To sensitize the paper dissolve the ferric salt and the ferrocyanide in water, separately, as they are then not sensitive to the action of light. The solutions should be *mixed* and *applied* to the paper only in a *dark room*.

Although there is the highest authority for "floating the paper to be sensitized for two minutes on the surface of the liquid" yet the best American practice is to apply the solution with a soft flat brush about four inches wide. The main object is to obtain an even coat, which may usually

---

[*] Published by the D. Van Nostrand Company, New York. † American Edition revised and published by Edward L. Wilson, New York.

98   *THEORETICAL AND PRACTICAL GRAPHICS.*

be secured by a primary coat of horizontal strokes followed by an overlay of vertical strokes; the second coat applied before the first dries. If necessary, another coat of diagonal strokes may be given to secure evenness. The thicker the coating given the longer the time required in printing. A bowl or flat dish or plate will be found convenient for holding the small portion of the solution required for use at any one time. The chemicals should not get on the back of the sheet.

Each sheet, as coated, should be set in a dark place to dry, either "tacked to a board by two adjacent corners", or "hung on a rack or over a rod", or "placed in a drawer—one sheet in a drawer",—varying instructions, illustrating the quite general truth that there are usually several almost equally good ways of doing a thing.

273. To copy a drawing place the prepared paper, sensitized side up, on a drawing-board or printing-frame on which there has been fastened, *smoothly*, either a felt pad or canton flannel cloth. The drawing is then immediately placed over the first sheet, inked side up, and contact secured between the two by a large sheet of plate glass, placed over all.

Exposure in the direct rays of the sun for four or five minutes is usually sufficient. The progress of the chemical action can be observed by allowing a corner of the paper to project beyond the glass. It has a grayish hue when sufficiently exposed.

If the sun's rays are not direct, or if the day is cloudy, a proportionately longer time is required, running up in the latter case, from minutes into hours. Only experiment will show whether one's solution is "quick" or "slow;" or the time required by the degree of cloudiness.

A solution will print more quickly if the amount of water in it be increased or if more iron is used; but in the former case the print will not be as dark, while in the latter the results, as to whiteness of lines, are not so apt to be satisfactory.

Although fair results can be obtained with paper a month or more after it has been sensitized, yet they are far more satisfactory if the paper is prepared each time (and dried) just before using.

On taking the print out of the frame it should be immediately immersed and thoroughly washed in *cold water* for from three to ten minutes, after which it may be dried in either of the ways previously suggested.

If many prints are being made, the water should be frequently changed so as not to become charged with the solution.

274. The entire process, while exceedingly simple in theory, varies, as to its results, with the experience and judgment of the manipulator. To his choice the decision is left between the following standard recipes for preparing the sensitizing solution. The "parts" given are all by weight. In every case the potash should be pulverized, to facilitate its dissolving.

No. 1.   (From *Le Génie Civil*).

Solution No. 1. { Red Prussiate of Potash... 8 parts.
                   { Water... 70 parts.

Solution No. 2. { Citrate of Iron and Ammonia .... 10 parts.
                   { Water... 70 parts.

Filter the solutions separately, mix equal quantities and then filter again.

No. 2.   (From U. S. Laboratory at Willett's Point).

Solution No. 1. { Double Citrate of Iron and Ammonia..................... 1 ounce.
                   { Water... 4 ounces.

Solution No. 2. { Red Prussiate of Potassium .... 1 ounce.
                   { Water... 4 ounces.

BLUE-PRINT PROCESS.

No. 3. (Lietze's Method).

Stock Solution. { 5 ounces, avoirdupois, Red Prussiate of Potash.
{ 32 fluid ounces ....Water.

"After the red prussiate of potash has been dissolved—which requires from one to two days—the liquid is filtered. This solution remains in good condition for a long time. Whenever it is required to sensitize paper, dissolve, for every two hundred and forty square feet of paper

{ 1 ounce, avoirdupois, Citrate of Iron and Ammonia,
{ 4½ fluid ounces ....Water,

and mix this with an equal volume of the stock solution.

The reason for making a stock solution of the red prussiate of potash is, that it takes a considerable time to dissolve and because it must be filtered. There are many impurities in this chemical which can be removed by filtering. Without filtering, the solution will not look clear. The reason for making no stock solution of the ferric citrate of ammonia is that such solution soon becomes moldy and unfit for use. This ferric salt is brought into the market in a very pure state and does not need to be filtered after being dissolved. It dissolves very rapidly. In the solid form it may be preserved for an unlimited time, if kept in a well-stoppered bottle and protected against the moisture of the atmosphere. A solution of this salt, or a mixture of it with the solution of red prussiate of potash, will remain in a serviceable condition for a number of days, but it will spoil, sooner or later, according to atmospheric conditions. . . . Four ounces of sensitizing solution, for blue prints, are amply sufficient for coating one hundred square feet of paper, and cost about six cents."

For copying tracings in *blue* lines or *black*, on a white ground, one may either employ the recipes given in Lietze's and Pettit's works or obtain paper, already sensitized, from the leading dealers in draughtsmen's supplies. The latter course has become quite as economical, also, for the ordinary blue-print, as the preparing of one's own supply.

For copying a drawing in *any* desired color the following method, known as *Tilhet's*, is said to give good results: "The paper on which the copy is to appear is first dipped in a bath consisting of 30 parts of white soap, 30 parts of alum, 40 parts of English glue, 10 parts of albumen, 2 parts of glacial acetic acid, 10 parts of alcohol of 60°, and 500 parts of water. It is afterward put into a second bath, which contains 50 parts of burnt umber ground in alcohol, 20 parts of lampblack, 10 parts of English glue, and 10 parts of bichromate of potash in 500 parts of water. They are now sensitive to light, and must, therefore, be preserved in the dark. In preparing paper to make the positive print another bath is made just like the first one, except that lampblack is substituted for the burnt umber. To obtain colored positives the black is replaced by some red, blue or other pigment.

In making the copy the drawing to be copied is put in a photographic printing frame, and the negative paper laid on it, and then exposed in the usual manner. In clear weather an illumination of two minutes will suffice. After the exposure the negative is put in water to develop it, and the drawing will appear in white on a dark ground; in other words, it is a negative or reversed picture. The paper is then dried and a positive made from it by placing it on the glass of a printing-frame and laying the positive paper upon it and exposing as before. After placing the frame in the sun for two minutes the positive is taken out and put in water. The black dissolves off without the necessity of moving back and forth."

PHOTO- AND OTHER PROCESSES.

275. If a drawing is to be reproduced on a different scale from that of the original, some one of the processes which admits of the use of the camera is usually employed. Those of most importance to the draughtsman are (1) *wood engraving;* (2) the "wax process" or *cerography;* (3) *lithography,* and (4) the various methods in which the photographic negative is made on a film of gelatine which is then used *directly*—to print from, or *indirectly*—in obtaining a metal plate from which the impressions are taken.

In the first three named above the use of the camera is not invariably an element of the process.

All under the fourth head are essentially photo-processes and their already large number is constantly increasing. Among them may be mentioned *photogravure, collotype, phototype, autotype, photoglyph, albertype, heliotype,* and *heliogravure.*

WOOD ENGRAVING.

276. There is probably no process that surpasses the best work of skilled engravers on wood. This statement will be sustained by a glance at Figs. 14, 15, 20-24, 134, 136, and those illustrating mathematical surfaces, in the next chapter. Its expensiveness and the time required to make an illustration by this method are its only disadvantages.

Although the camera is often employed to transfer the drawing to the boxwood block in which the lines are to be cut, yet the original drawing is quite as frequently made *in reverse,* directly on the block, by a professional draughtsman who is supposed to have at his disposal either the object to be drawn or a photograph or drawing thereof. The outlines are pencilled on the block and the shades and shadows given in brush tints of India ink, re-enforced, in some cases, by the pencil, for the deepest shadows.

The "high lights" are brought out by Chinese white. A medium wash of the latter is also usually spread upon the block as a general preliminary to outlining and shading.

The task of the engraver is to reproduce faithfully the most delicate as well as the strongest effects obtained on the block with pencil and brush, cutting away all that is not to appear in black in the print. The finished block may then be used to print from directly, or an electrotype block can be obtained from it which will stand a large number of impressions much better than the wood.

CEROGRAPHY.

277. For map-making, illustrations of machinery, geometrical diagrams and all work mainly in straight lines or simple curves, and not involving too delicate gradations, the cerographic or "wax process" is much employed. For clearness it is scarcely surpassed by steel engraving. Figures 36, 90 and 107 are good specimens of the effects obtainable by this method. The successive steps in the process are (a) the laying of a thin, even coat of wax over a copper plate; (b) the transfer of the drawing to the surface of the wax, either by tracing or—more generally—by photography; (c) the re-drawing or rather the cutting of these lines in the wax, the stylus removing the latter to the surface of the copper; (d) the taking of an electrotype from the plate and wax, the deposit of copper filling in the lines from which the wax was removed.

Although in the preparation of the original drawing the lines may preferably be inked yet this is not absolutely necessary, provided a pencil of medium grade be employed.

Any letters desired on the final plate may be also pencilled in their proper places, as the engraver makes them on the wax with type.

A surface on which section-lining or cross-hatching is desired may have that fact indicated upon it in writing, the direction and number of lines to the inch being given. Such work is then done with a ruling machine.

Errors may readily be corrected, as the surface of the wax may be made smooth, for recutting, by passing a hot iron over it.

LITHOGRAPHY.—PHOTO-LITHOGRAPHY.—CHROMO-LITHOGRAPHY.

278. For the lithographic process a fine-grained, imported limestone is used. The drawing is made with a greasy ink—known as "lithographic"—upon a specially prepared paper, from which it is transferred, under pressure, to the surface of the stone. The un-inked parts of the stone are kept thoroughly moistened with water, which prevents the printer's ink (owing to the grease which the latter contains) from adhering to any portion except that from which the impressions are desired.

Photo-lithography is simply lithography, with the camera as an adjunct. The positive might be made directly upon the surface of the stone by coating the latter with a sensitizing solution; but, in general, for convenience, a sensitized gelatine film is exposed under the negative, and by subsequent treatment gives an image in relief which, after inking, can be transferred to the surface of the stone as in the ordinary process.

Chromo-lithography, or lithography in colors, has been a very expensive process owing to its requiring a separate stone for each color. Recent inventions render it probable that it will be much simplified and the expense correspondingly reduced. The details of manipulation are closely analogous to those for ink prints.

When colored plates are wanted, in which delicate gradations shall be indicated, chromo-lithography may preferably be adopted; although "half tones," with colored inks, give a scarcely less pleasing effect, as illustrated by Figs. 7–10, Plate II. But for simple line-work, in two or more colors, one may preferably employ either cerography or photo-engraving, each of which has not only an advantage, as to expense, over any lithographic process, but also this in addition—that the blocks can be used by any printer; whereas lithographing establishments necessarily not only prepare the stone but also do the printing.

PHOTO-ENGRAVING.—PHOTO-ZINCOGRAPHY.

279. In this popular and rapid process a sensitized solution is spread upon a smooth sheet of zinc and over this the photographic negative is placed. Where not acted on by the light the coating remains soluble and is washed away, exposing the metal, which is then further acted on by acids to give more *relief* to the remaining portions.

Except as described in Art. 281 this process is only adapted to inked work in lines or dots, which it reproduces faithfully, to the smallest detail. Among the best photo-engravings in this book are Figs. 12, 13, 50, 79 and 80.

280. The following instructions for the preparation of drawings, for reproduction by this process, are those of the American Society of Mechanical Engineers as to the illustration of papers by its members, and are, in general, such as all the engraving companies furnish on application.

"All lines, letters and figures must be *perfectly black* on a white ground. Blue prints are not available, and red figures and lines will not appear. The smoother the paper, and the blacker the ink, the better are the results. Tracing-cloth or paper answers very well, but rough paper—even

Whatman's—gives bad lines. India ink, ground or in solution, should be used; and the best lines are made on Bristol board, or its equivalent with an enameled surface. Brush work, in tint or grading, unfits a drawing for immediate use, since only line work can be photographed. Hatching for sections need not be completed in the originals, as it can be done easily by machine on the block. If draughtsmen will indicate their sections unmistakably, they will be properly lined, and tints and shadows will be similarly treated.

The best results may be expected by using an original twice the height and width of the proposed block. The reduction can be greater, provided care has been taken to have the lines far enough apart, so as not to mass them together. Lines in the plate may run from 70 to 100 to the inch, and there should be but half as many in a drawing which is to be reduced one half; other reductions will be in like proportion.

Draughtsmen may use photographic prints from the objects if they will go over with a carbon ink all the lines which they wish reproduced. The photographic color can be bleached away by flowing a solution of bi-chloride of mercury in alcohol over the print, leaving the pen lines only. Use half an ounce of the salt to a pint of alcohol.

Finally, lettering and figures are most satisfactorily printed from type. Draughtsmen's best efforts are usually thus excelled. Such letters and figures had therefore best be left in pencil on the drawings, so they will not photograph but may serve to show what type should be inserted."

To the above hints should be added a caution as to the use of the rubber. It is likely to diminish the intensity of lines already made and to affect their sharpness; also to make it more difficult to draw clear-cut lines wherever it has been used.

It may be remarked with regard to the foregoing instructions that they aim at securing that uniformity, as to general appearance, which is usually quite an object in illustration. But where the preservation of the individuality and general characteristics of one's work is of any importance whatever, the draughtsman is advised to letter his own drawings and in fact finish them entirely, himself, with, perhaps, the single exception of section-lining, which may be quickly done by means of *Day's Rapid Shading Mediums* or by other technical processes.

281. *Half Tones.* Photo-zincography may be employed for reproducing delicate gradations of light and shade, by breaking up the latter when making the photographic negative. The result is called a *half tone* and it is one of the favorite processes for high-grade illustration. Figs. 95 and 130 illustrate the effects it gives. On close inspection a series of fine dots in regular order will be noticed, so that no tone exists unbroken, but all have more or less white in them.

The methods of breaking up a tone are very numerous. The first patent dates back to 1852. The principle is practically the same in all, viz., between the object to be photographed and the plate on which the negative is to made there is interposed a "screen" or sheet of thin glass on which the desired mesh has been previously photographed.

In the making of the "screen" lies the main difference between the variously-named methods. In Meissenbach's method, by which Figs. 95 and 130 were made, a photograph is first taken, on the "screen," of a pane of clear glass in which a system of parallel lines—one hundred and fifty to the inch—has been cut with a diamond. The ruled glass is then turned at right angles to its first position and its lines photographed on the screen over the first set, the times of exposure differing slightly in the two cases, being generally about as 2 to 3.

This process is well adapted to the reproduction of "wash" or brush-tinted drawings, photographs, etc. The object to be represented, if small, may preferably be furnished to the engraving company and they will photograph it direct.

GELATINE FILM PHOTO-PROCESSES.

282. As stated in Art. 275, in which a few of the above processes are named, a gelatine film may be employed, either as an adjunct in a method resulting in a metal block, or to print from directly; in the latter case the prints must be made, on special paper, by the company preparing the film. In the composition and manipulation of the film lies the main difference between otherwise closely analogous processes. For any of them the company should be supplied with either the original object or a good drawing or photographic negative thereof.

Not to unduly prolong this chapter—which any intelligible distinction between the various methods would involve, yet to give an idea of the general principles of a gelatine process I conclude with the details of the preparation of a *heliotype* plate, given in the language of one company's circular. Figs. 1—5 of Plate II illustrate the effect obtained by it.

"Ordinary cooking gelatine forms the basis of the positive plate, the other ingredients being bichromate of potash and chrome alum. It is a peculiarity of gelatine, in its normal condition, that it will absorb *cold water*, and swell or expand under its influence, but that it will dissolve in *hot water*. In the preparation of the plate, therefore, the three ingredients just named, being combined in suitable proportions, are dissolved in hot water, and the solution is poured upon a level plate of glass or metal, and left there to dry. When dry it is about as thick as an ordinary sheet of parchment, and is stripped from the drying-plate, and placed in contact with the previously-prepared negative, and the two together are exposed to the light. The presence of the bichromate of potash renders the gelatine sheet sensitive to the action of light; and wherever light reaches it, the plate, which was at first gelatinous or absorbent of water, becomes leathery or waterproof. In other words, wherever light reaches the plate, it produces in it a change similar to that which tanning produces upon hides in converting them into leather. Now it must be understood that the negative is made up of transparent parts and opaque parts; the transparent parts admitting the passage of light through them, and the opaque parts excluding it. When the gelatine plate and the negative are placed in contact, they are exposed to light with the negative *uppermost*, so that the light acts through the translucent portions, and waterproofs the gelatine underneath them; while the opaque portions of the negative shield the gelatine underneath them from the light, and consequently those parts of the plate remain unaltered in character. The result is a thin, flexible sheet of gelatine, of which a portion is waterproofed, and the other portion is absorbent of water, the waterproofed portion being the image which we wish to reproduce. Now we all know the repulsion which exists between water and any form of *grease*. Printer's ink is merely grease united with a coloring-matter. It follows, that our gelatine sheet, having water applied to it, will absorb the water in its unchanged parts; and, if ink is then rolled over it, the ink will adhere only to the waterproofed or altered parts. This flexible sheet of gelatine, then, prepared as we have seen, and having had the image impressed upon it, becomes the *heliotype plate*, capable of being attached to the bed of an ordinary printing-press, and printed in the ordinary manner. Of course, such a sheet must have a solid base given to it, which will hold it firmly on the bed of the press while printing. This is accomplished by uniting it, under water, with a metallic plate, exhausting the air between the two surfaces, and attaching them by atmospheric pressure. The plate, with the printing surface of gelatine attached, is then placed on an ordinary platen printing-press, and inked up with ordinary ink. A mask of paper is used to secure white margins for the prints; and the impression is then made, and is ready for issue."

"*The study of Descriptive Geometry possesses an important philosophical peculiarity, quite independent of its high industrial utility. This is the advantage which it so pre-eminently offers in habituating the mind to consider very complicated geometrical combinations in space, and to follow with precision their continual correspondence with the figures which are actually traced—of thus exercising to the utmost, in the most certain and precise manner, that important faculty of the human mind which is properly called 'imagination,' and which consists, in its elementary and positive acceptation, in representing to ourselves, clearly and easily, a vast and variable collection of ideal objects, as if they were really before us. . . . . . While it belongs to the geometry of the ancients by the character of its solutions, on the other hand it approaches the geometry of the moderns by the nature of the questions which compose it. These questions are in fact eminently remarkable for that generality which constitutes the true fundamental character of modern geometry; for the methods used are always conceived as applicable to any figures whatever, the peculiarity of each having only a purely secondary influence.*"
                   AUGUSTE COMTE: Cours de Philosophie Positive.

"*A mathematical problem may usually be attacked by what is termed in military parlance the method of 'systematic approach;' that is to say, its solution may be gradually felt for, even though the successive steps leading to that solution cannot be clearly foreseen. But a Descriptive Geometry problem must be seen through and through before it can be attempted. The entire scope of its conditions as well as each step toward its solution must be grasped by the imagination. It must be 'taken by assault.'*"
            GEORGE SYDENHAM CLARKE, Captain, Royal Engineers.

CHAPTER X.

PROJECTIONS AND INTERSECTIONS BY THE THIRD-ANGLE METHOD.—THE DEVELOPMENT OF SURFACES FOR SHEET METAL PATTERN MAKING.—PROJECTIONS, INTERSECTIONS AND TANGENCIES OF DEVELOPABLE, WARPED AND DOUBLE CURVED SURFACES, BY THE FIRST-ANGLE METHOD.

383. The mechanical drawings preliminary to the construction of machinery, blast furnaces, stone arches, buildings, and, in fact, all architectural and engineering projects, are made in accordance with the principles of Descriptive Geometry. When fully dimensioned they are called *working drawings*.

The object to be represented is supposed to be placed in either the first or the third of the four angles formed by the intersection of a horizontal plane, H, with a vertical plane, V. (Fig. 228).

The representations of the object upon the planes are, in mathematical language, *projections*,* and are obtained by drawing perpendiculars to the planes H and V from the various points of the object, the point of intersection of each such projecting line with a plane giving a *projection* of the original point. Such drawings are, obviously, not "views" in the ordinary sense, as they lack the perspective effect which is involved in having the point of sight at a finite distance; yet in ordinary parlance the terms *top view*, *horizontal projection* and *plan* are used synonymously; as are *front view* and *front elevation* with *vertical projection*, and *side elevation* with *profile view*, the latter on a plane perpendicular to both H and V and called the *profile plane*.

Until the last decade of the first century of Descriptive Geometry (1795–1895) problems were solved as far as possible in the first angle. As the location of the object in the third angle—that is, below the horizontal plane and behind the vertical—results in a grouping of the views which is in a measure self-interpreting, the *Third Angle Method* is, however, to a considerable degree supplanting the other for machine-shop work.

The advantageous grouping of the projections which constitutes the only—though a quite sufficient—justification for giving it special treatment, is this: The front view being always the *central* one of the group, the top view is found *at the top;* the view of the right side of the object appears *on the right;* of the left-hand side *on the left,* etc. Thus, in Fig. 228 (a), with the hollow block $BDFS$ as the object to be represented, we have $adcs$ for its horizontal projection, $c'd'e'f'$ for its vertical projection, $f''e''s''s''$ for the side elevation; then on rotating the plane H clockwise on G.L. into coincidence with V, and the profile plane P about QR until the projection $f''e''s''s''$ reaches $f'''e'''s'''s'''$, we would have that location of the views which has just been described.

The lettering shows that each projection represents that side of the object which is toward the plane of projection.

384. The same grouping can be arrived at by a different conception, which will, to some, have advantages over the other. It is illustrated by Fig. 228 (b), in which the same object as before is

---
*For the convenience of those who have to take up this subject without previous study of Descriptive Geometry the Third-Angle section of this chapter is made complete in itself, by the re-statement of the principles involved and which have been treated at somewhat greater length in the previous chapter; although a review of such matter may be by no means disadvantageous to those who have already been over the fundamentals.

supposed to be surrounded by a system of mutually-perpendicular transparent planes, or, in other words, to be in a box having glass sides, and on each side a drawing made of what is seen through that side, excluding the idea, as before, of perspective view, and representing each point by a per-

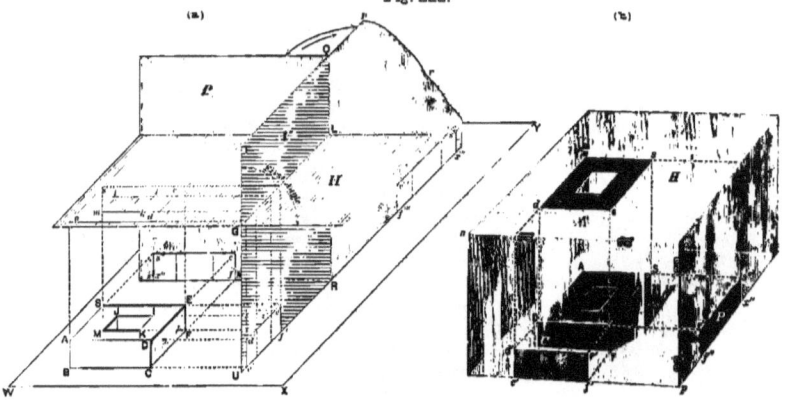

Fig. 228.

pendicular from it to the plane. The whole system of box and planes, in the wood-cut, is rotated 90° from the position shown in Fig. 228 (a), bringing them into the usual position, in which the observer is looking perpendicularly toward the vertical plane.

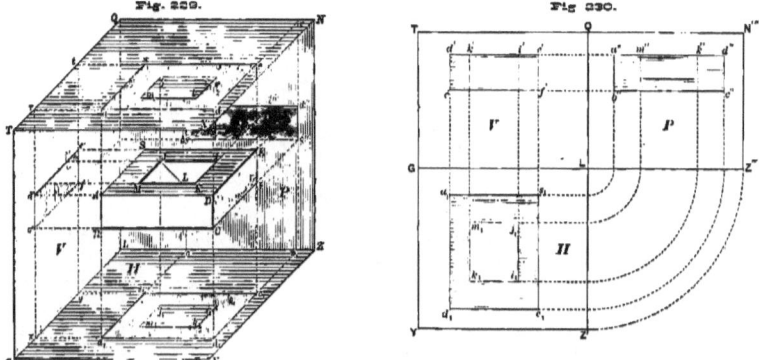

385. In Fig. 229 we may illustrate either the First or the Third Angle method, as to the top view of the object; $ades$ in the upper plane being the *plan* by the latter method, and $a_1 d_1 e_1 s_1$ by the former.

ORTHOGRAPHIC PROJECTION OF SOLIDS.

Disregarding $QTXN$ we have the object and planes illustrating the first-angle method throughout, the lettering of each projection showing that it represents the side of the object *farthest* from the plane, making it the exact reverse of the third-angle system.

In the ordinary representation the same object would be represented simply by its three views as in Fig. 230. In the elevations the short-dash lines indicate the invisible edges of the hole.

The arcs show the rotation which carries the profile view into its proper place.

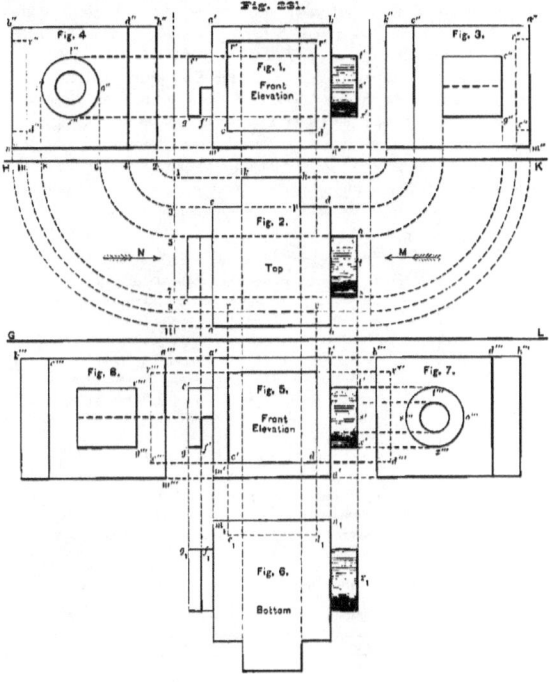

Fig. 231.

386. For the sake of more readily contrasting the two methods a group of views is shown in Fig. 231, all above GL, illustrating an object by the First Angle system, while all below HK represents the same object by the Third Angle method.

When looking at Figures 1, 2, 3 and 4 the observer queries: What is the object, *in space*, whose *front* is like Fig. 1, *top* is like Fig. 2, *left* side is like Fig. 3 and *right* side like Fig. 4?

For the view of the left side he might imagine himself as having been at first between G and H, looking in the direction of arrow N, after which both himself and the object were turned, together to the right, through a ninety-degree arc, when the same side would be presented to his view in Fig. 3. Similarly, looking in the direction of the arrow M, an equal rotation to the left, as indicated by the arcs 1-2, 3-4, 5-6, etc., would give in Fig. 4 the view obtained from direction M. His mental queries would then be answered about as follows: Evidently a cubical block with a rectangular recess—$c'c'd'e'$—in front; on the rear a prismatic projection, of thickness ph and whose height equals that of the cube; a short cylindrical ring projecting from the right face of the cube; an angular projecting piece on the left face.

In Fig. 2 the line cc is in short dashes, as in that view the back plane of the recess $c'c'd'e'$ would be invisible. In Fig. 4 the back plane of the same recess is given the letters, $c''d''$, of the edge nearest the observer from direction M.

To illustrate the third angle method by Fig. 231 we ignore all above the line HK. In Fig. 5 we have the same front elevation as before, but *above* it the view of the *top*; *below* it the view of the *bottom* exactly as it would appear were the object held before one as in Fig. 5, then given a ninety-degree turn, around $a'b'$, until the under side became the front elevation.

Fig. 7 may as readily be imagined to be obtained by a shifting of the object as by the rotation of a plane of projection; for by translating the object to the right, from its position in Fig. 5, then rotating it to the left 90° about $b'a'$, its right side would appear as shown.

387. For convenient reference a general resumé of terms, abbreviations and instructions is next presented, once for all, for use in both the Third Angle and First Angle methods.

(1) H, V, P . . . . the *horizontal, vertical* and *profile* planes of projection respectively.
(2) H-projector . . . . the *projecting line* which gives the *horizontal projection* of a point.
(3) V-projector . . . . the projecting line giving the projection of a point on V.
(4) Projector-plane . . . . the profile plane containing the projectors of a point.
(5) h. p. . . . . the *horizontal projection* or *plan* of a point or figure.
(6) v. p. . . . . the *vertical projection* or *elevation* of a point or figure.
(7) h. t. . . . . *horizontal trace,* the intersection of a line or surface with H.
(8) v. t. . . . . *vertical trace,* the intersection of a line or surface with V.
(9) H-traces, V-traces . . . . plural of horizontal and vertical traces respectively.
(10) G. L. . . . . *ground line,* the line of intersection of V and H.
(11) V-parallel . . . . a line parallel to V and lying in a given plane.
(12) A horizontal . . . . any horizontal line lying in a given plane.
(13) Line of declivity . . . . the steepest line, with respect to one plane, that can lie in another plane.
(14) Rabatment . . . . revolution into H or V about an axis in such plane.
(15) Counter-rabatment or revolution . restoration to original position.

388.   *For Problems relating solely to the Point, Line and Plane.*

Given lines should be fine, continuous, black; *required lines* heavy, continuous, black or red; *construction lines* in fine, continuous red, or short-dash black; *traces of an auxiliary plane,* or *invisible traces* of any plane, in dash-and-three-dot lines. ——— · · · ——— · · · ——— · · · ——— · · ·

For Problems relating to Solid Objects.

(1) *Pencilling.* Exact; generally completed for the whole drawing before any inking is done; the work usually from centre lines, and from the larger — and nearer — parts of the object to the smaller or more remote.

(2) *Inking of the Object.* Curves to be drawn before their tangents; fine lines uniform and drawn before the shade lines; shade lines next and with one setting of the pen, to ensure uniformity. On *tapering shade lines* see Art. 111.

(3) *Shade Lines.* In architectural *work* these would be drawn in accordance with a given direction of light.

In *American machine-shop practice* the *right-hand* and *lower* edges of a *plane* surface are made shade lines if they separate it from *invisible* surfaces. Indicate *curvature* by *line-shading* if not otherwise sufficiently evident. (See Fig. 288).

THE CONSTRUCTION AND FINISH OF WORKING DRAWINGS. 135

(4) *Invisible lines of the object*, black, invariably, in dashes nearly one-tenth of an inch in length. ----------
(5) *Inking of lines other than of the object.* When no colors are to be employed the following directions as to kind of line are those most frequently made. The lines may preferably be drawn in the order mentioned.
Centre lines, an alternation of dash and two dots. — ·· — ·· — ·· —
Dimension lines, a dash and dot alternately, with opening left for the dimension. — · — · —
Extension lines, for dimensions placed outside the views, in dash-and-dot as for a dimension line. — · — · —
Ground line, (when it cannot be advantageously omitted) a continuous heavy line. ——————
Construction and other explanatory lines in short dashes. - - - - - - - - - - - - - - - - - - -

(6) *When using colors* the centre, dimension and extension lines may be fine, continuous, red; or the former may be *blue*, if preferred. *Construction lines* may also be red, in short dashes or in fine continuous lines.
Instead of using bottled inks the carmine and blue may preferably be taken directly from Winsor and Newton cakes, "moist colors." Ink ground from the cake is also preferable to bottled ink.
Drawings of developable and warped surfaces are much more effective if their elements are drawn in some color.

(7) *Dimensions and Arrow-Tips*. The dimensions should invariably be in *black*, printed *free-hand* with a writing-pen, and *should read in line with the dimension line they are on*. On the drawing as a whole the dimensions should read either from the bottom or right-hand side. Fractions should have a *horizontal* dividing line; although there is high sanction for the omission of the dividing line, particularly in a mixed number.
Extended Gothic, Roman, Italic Roman and Reinhardt's form of Condensed Italic Gothic are the best and most generally used types for dimensioning.
The *arrow-tips* are to be always drawn *free-hand*, in *black*; to touch the lines between which they give a distance; and to make an acute instead of a right angle at their point.

389. *Working drawing of a right pyramid; base, an equilateral triangle 0.9" on a side; altitude, x.*
Draw first the equilateral triangle abc for the plan of the base, making its sides of the prescribed length. If we make the edge ab perpendicular to the profile plane, $O1$, the face eab will then appear in profile view as the straight line $e''b''$. Being a *right* pyramid, with a *regular* base, we shall find v, the plan of the vertex, equally distant from a, b and c; and va, vb, vc for the plans of the edges.

Fig. 232.

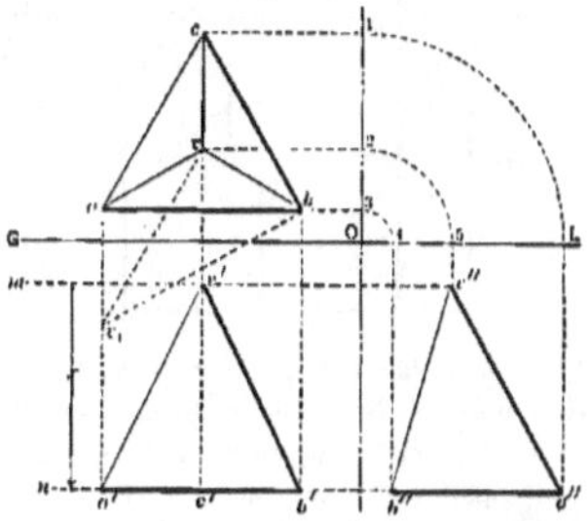

Parallel to G.L. and at a distance apart equal to the assigned height, x, draw mv' and uv'' as upper and lower limits of the front and side elevations; then, as the h.p. and v.p. of a point are always in the same perpendicular to G.L., we project v, a, b and c to their respective levels by the construction lines shown, obtaining $v'.a'b'c'$ for the *front elevation*.

Projectors to the profile plane from the points of the plan give 1, 2, 3, which are then carried in arcs about O, to L, 5, 4, and projected to their proper levels, giving the *side elevation*, $v''b''c''$.
As *the actual length of an edge* is not shown in either of the three views, we employ the following construction to ascertain it: Draw cv, perpendicular to vb, and make it equal to x; v_1b is then the real length of the edge, shown by rabatment about vb.

Fig. 233.

The *development* of the pyramid (Fig. 233) may be obtained by drawing an arc $ABCA_1$ of radius $= v_1b$ (the true length of edge, from Fig. 232) and on it laying off the chords AB, BC, CA_1 equal to ab, bc, ca of the plan; then $VABCA_1$ is the plane area which, folded on VC and VB, would give a model of the pyramid represented.

136   THEORETICAL AND PRACTICAL GRAPHICS.

390. *Working drawing of a semi-cylindrical pipe:* outer diameter, x; inner diameter, y; height, z. For the *plan* draw concentric semi-circles $a e d$ and $b s c$, of diameters x and y respectively, joining their extremities by straight lines $a b$, $c d$. At a distance apart of z inches draw the upper and lower limits of the elevations, and project to these levels from the points of the plan.

In the side view the thickness of the shell of the cylinder is shown by the distance between $c''f''$ and $s''t''$—the latter so drawn as to indicate an invisible limit or line of the object.

The line shading would usually be omitted, the shade lines generally sufficing to convey a clear idea of the form.

391. *Half of a hollow, hexagonal prism.* In a semi-circle of diameter $a d$ step off the radius three times as a chord, giving the vertices of the plan $a b c d$ of the outer surface. Parallel to $b c$, and at a distance from it equal to the assigned thickness of the prism, draw $e f$, terminating it on lines (not shown) drawn through b and c at 60° to $a d$. From e and f draw $e h$ and $f g$, parallel respectively to $a b$ and $c d$. Drawing $a'e''$ and $m't''$ as upper and lower limits, project to them as in preceding problems for the front and side elevations.

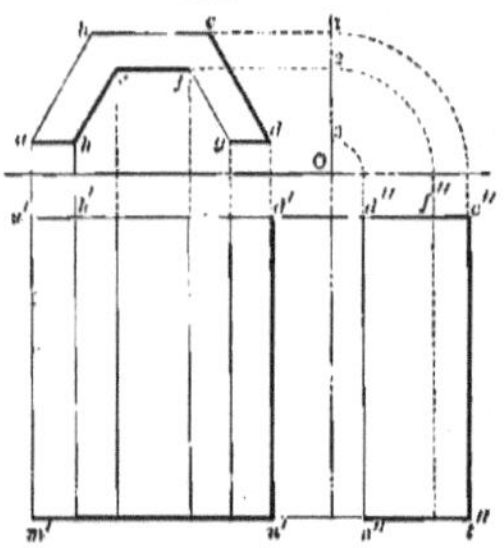

392. *Working drawing of a hollow, prismatic block, standing obliquely to the vertical and profile planes.*

Let the block be $2'' \times 3'' \times 1''$ outside, with a square opening $1'' \times 1'' \times 1''$ through it in the direction of its thickness. Assuming that it has been required that the two-inch edges should be vertical, we first draw, in Fig. 236, the plan $a s x b$, $3'' \times 1''$, on a scale of 1:2. The inch-wide opening through the centre is indicated by the short-dash lines.

For the elevations the upper and lower limits are drawn $2''$ apart, and a, b, s, x, etc., projected to them. The elevations of the opening are between levels $m'n''$ and $k'k''$, one inch apart and equi-distant from the upper and lower outlines of the views. The dotted construction lines and the lettering will enable the student to recognize the three views of any point without difficulty.

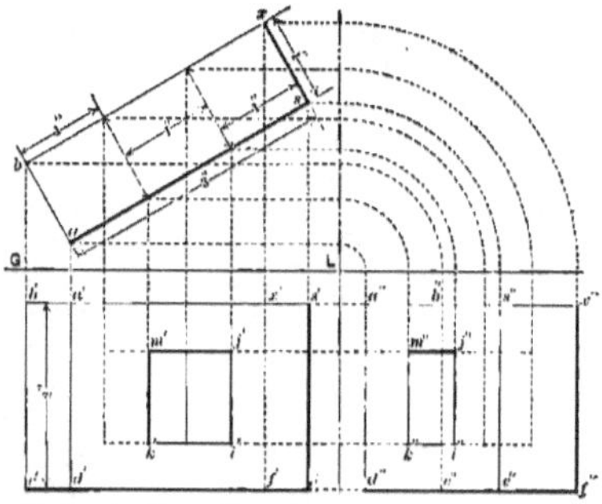

393. In Fig. 237 we have the same object as that illustrated by Fig. 236, but now represented as cut by a vertical plane whose horizontal trace is $r y$. The parts of the block that are actually cut by the plane are shown in section-lines in the elevations. This is done here and in some later examples merely to aid the beginner in understanding the views; but, *in engineering practice, section-lining is rarely done on views not perpendicular to the section plane.*

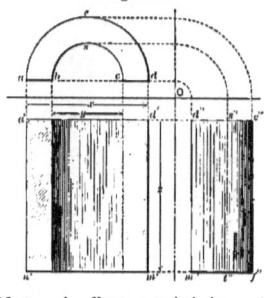

Fig. 234.

Fig. 235.

Fig. 236.

PROJECTION OF SOLIDS.—WORKING DRAWINGS.

394. *Suppression of the Ground Line.* In machine drawing it is customary to omit the ground line, since the *forms* of the various views—which alone concern us—are independent of the distance of the object from an imaginary horizontal or vertical plane. We have only to remember that all elevations of a point are at the same level; and that if a ground line or trace of any vertical plane is wanted, it will be perpendicular to the line joining the plan of a point with its projection on such vertical plane.

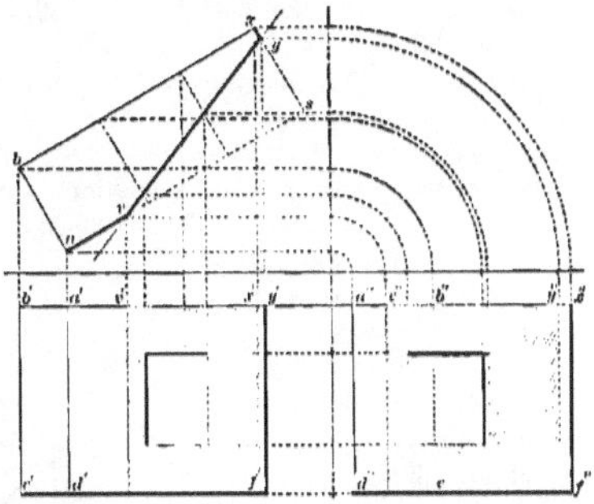

395. *Sections. Sectional Views.* Although earlier defined (Art. 70), a re-statement of the distinction between these terms may well precede problems in which they will be so frequently employed.

When a plane cuts a solid, that portion of the latter which comes *in actual contact* with the cutting plane is called the *section*.

A *sectional view* is a view *perpendicular to the cutting plane*, and showing not only the section but also the object itself as if seen through the plane. When the cutting plane is *vertical* such a view is called a *sectional elevation*; when *horizontal*, a *sectional plan*.

396. *Working drawing of a regular, pentagonal pyramid, hollow, truncated by an oblique plane; also the development, or "pattern," of the outer surface below the cutting plane.* For data take the altitude at $2''$; inclination of faces, $\theta°$ (meaning any arbitrary angle); inclination of section plane, $30°$; distance between inner and outer faces of pyramid, $\tfrac{1}{4}''$.

(1) Locate c and c' (Fig. 238) for the plan and elevation of the vertex, taking them sufficiently apart to avoid the overlapping of one view upon the other. Through c draw the horizontal line ST, regarding it not only as a centre line for the plan but also as the h. t. of a central, vertical, reference plane, parallel to the ordinary vertical plane of projection.

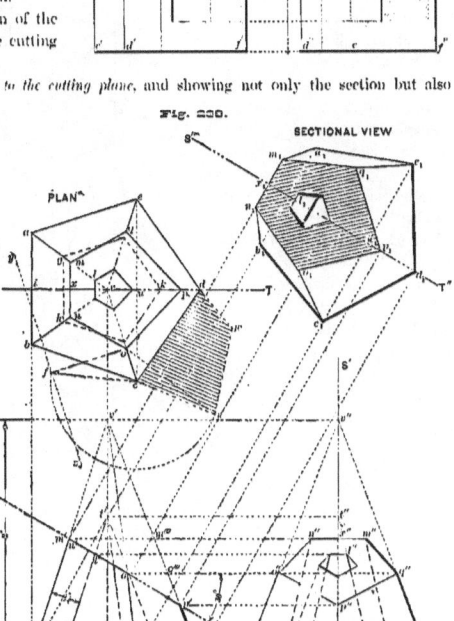

Fig. 237.

Fig. 238.

138   THEORETICAL AND PRACTICAL GRAPHICS.

(The student should note that for convenience Fig. 238 is repeated on this page.)
On the vertical line vv' (at first indefinite in length) lay off $v's'$ equal to $2''$, for the altitude (and axis) of the pyramid, and through s' draw an indefinite horizontal line, which will contain the v. p. of the base, in both front and side views.

Draw $v'b'$ at $\theta°$ to the horizontal. It will represent the v. p. of an outer *face* of the pyramid, and b' will be the v. p. of the edge ab of the *base*. The base $abcde$ is then a regular pentagon circumscribed about a circle of centre v and radius $vi = s'b'$. Since the angle avb is $72°$ (Art. 92) we get a starting corner, a or b, by drawing va or vb at $36°$ to ST, to intercept the vertical through b'. The plans of the edges of the pyramid are then va, vb, vc, vd and ve. Project d to d' and draw $v'd'$ for the elevation of vd; similarly for vc and ve, which happen in this case to coincide in vertical projection.

For the inner surface of the pyramid, whose faces are at a perpendicular distance of $\frac{1}{4}''$ from the outer, begin by drawing $g'l'$ parallel to and $\frac{1}{4}''$ from the face projected in $b'v'$; this will cut the axis at a point t' which will be the vertex of the inner surface, and $g't'$ will represent the elevation of the inner face that is parallel to the face $avb - v'b'$; while gh, vertically above g' and included between va and vb, will be the plan of the lower edge of this face. Complete the pentagon $gh\cdots k$ for the plan of the inner base; project the corners to $b'd'$ and join with t' to get the elevations of the interior edges.

The Section. In our figure let $G'H'$ be the section plane, situated perpendicular to the vertical plane and inclined $30°$ to the horizontal. It intersects $v'd'$ in p', which projects upon vd at p. Similarly, since $G'H'$ cuts the edges $v'c'$ and $v'e'$ at points projected in o', we project from the latter to vc and ve, obtaining o and q. A like construction gives m and n. The polygon $monpq$ is then the plan of the outer boundary of the section.

The inner edge $g't'$ is cut by the section plane at l', which projects to both vh and vg, giving the parallel to mn through l. The inner boundary of the section may then be completed either by determining all its vertices in the same way or on the principle that its sides will be parallel to those of the outer polygon, since any two planes are cut by a third in parallel lines.

The line $m'p'$ is the vertical projection of the entire section.

(2) *The side elevation.* This might be obtained exactly as in the five preceding figures, that is, by actually locating the side vertical, or *profile,* plane, projecting upon it and rotating through an arc of 90°. In engineering practice, however, the method now to be described is in far more general use. It does not do away with the profile plane, on the contrary presupposes its existence, but instead of actually locating it and drawing the arcs which so far have kept the relation of the views constantly before the eye, it reaches the same result in the following manner: A vertical line $S'T'$ is drawn at some convenient distance to the right of the front elevation; the distance, from ST, of any point of the plan, is then laid off horizontally from $S'T'$, at the same height as the front elevation of the point. For, as earlier stated, ST was to be regarded as the horizontal trace of a vertical plane. Such plane would, evidently, cut a *profile* plane in a vertical line, which we may call $S'T'$, and let the $S'T'$ of our figure represent it after a ninety-degree rotation has occurred. The distances of all points of the object, to either the *front* or *rear* of the vertical plane on ST, would, obviously, be now seen as distances to the *left* or *right,* respectively, of the trace $S'T'$, and would be directly transferred with the dividers to the lines indicating their level. Thus, c'' is on the *level* of c', but is to the *right* of $S'T'$ the same distance that c is *above* (or, in reality, *behind*) the plane ST; that is, $c''d''$ equals cv. Similarly $d''b''$ equals ib; $u''x''$ equals ux.

It is usual, where the object is at all symmetrical, to locate these reference planes *centrally,* so that their traces, used as indicated, may *bisect* as many lines as possible, to make one setting of the dividers do double work.

(3) *True size of the Section. Sectional View.* If the section plane $G'H'$ were rotated directly about its trace on the central, vertical plane ST, until parallel to the paper, it would show the section $m'p'-mnup q$ in its true size; but such a construction would cause a confusion of lines, the new figure overlapping the front elevation. If, however, we transfer the plane $G'H'$—keeping it parallel to its first position during the motion—to some new position $S''T''$, and then turn it 90° on that line, we get $m_1 n_1 u_1 p_1 q_1$, the desired view of the section. The distances of the vertices of the section from $S''T''$ are derived from reference to ST exactly as were those in the side elevation; that is, $u_1 x_1 = m x = m''x''$. We thus see that one central, vertical, reference plane ST is auxiliary to the construction of two important views; $S'T'$ represents its intersection with the profile or side vertical plane, while $S''T''$ is its (transferred) trace upon the section plane $G'H'$. For the remainder of the sectional views the points are obtained exactly as above described for the section; thus $c'c_1 c_1$ is perpendicular to $S''T''$; $c_1 u_1$ equals cu, and $c_1 u_1$ equals cu.

(4) *To determine the actual length of the various edges.* The only edge of the original, uncut pyramid, that would require no construction in order to show its true length, is the extreme right-hand one, which—being parallel to the vertical plane, as shown by its plan vd being horizontal—is seen in elevation in its true size, $v'd'$. Since, however, all the edges of the pyramid are equal, we may find on $v'd'$ the true length of any *portion* of some other edge, as, for example $o'c'$, by taking that part of $v'd'$ which is intercepted between the same horizontals, viz.: $o'''d'$.

Were we compelled to find the true length of $o'c'$, oc, independently of any such convenient relation as that just indicated, we would apply one of the methods fully illustrated by Figs. 183, 184 and 187, or the following "shop" modification of one of them: Parallel to the plan oc draw a line yz, their distance apart to be equal to the *difference of level* of o' and c', which difference may be obtained from either of the elevations. From the plan o of the higher end of the line draw the common perpendicular of, and join f with c, obtaining the desired length fc.

(5) *To show the exact form of any face of the pyramid.* Taking, for example, the face $vcd p$, revolve op about the horizontal edge cd until it reaches the level of the latter. The actual distance

140   *THEORETICAL AND PRACTICAL GRAPHICS.*

of o from c, and of p from d will be the same after as before this revolution, while the paths of o and p during rotation will be projected in lines oe and pw, each perpendicular to cd; therefore, with c as a centre, cut the perpendicular oe by an arc of radius fc — just ascertained to be the real length of oc, and, similarly, cut pw by an arc of radius $dw = p'd'$; join e with c, w with d, draw wr and we have in $edwe$ the form desired.

(6) *The development of the outer surface of the truncated pyramid.* With any point V as a centre (Fig. 239) and with radius equal to the actual length of an edge of the pyramid (that is, equal to $c'd'$, Fig. 238), draw an indefinite arc, on which lay off the chords DC, CB, BA, AE, ED, equal respectively to the like-lettered edges of the base $abcde$; join the extremities of these chords with V: then on DV lay off $DP = d'p'$; make $CO = EQ = d'o''' =$ the real length of $e'o'$; also $BN = AM = d'm''' =$ the actual length of $a'm'$ and $b'n'$; join the points $P, O,$ etc., thus obtaining the development of the outer boundary of the section. The pattern $A_1 B_1 CD E_1$ of the base is obtained from the plan in Fig. 238, while $NMq_2p_2o_2$ is a duplicate of the shaded part of the *sectional view* in the same figure.

(7) *In making a model* of the pyramid the student should use heavy Bristol board, and make allowance, wherever needed, of an extra width for overlap, slit as at x, y and z (Fig. 239). On this

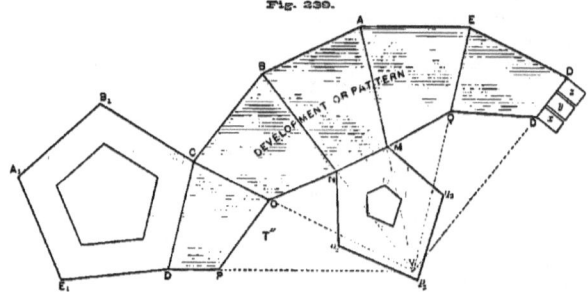

Fig. 239.

overlap put the mucilage which is to hold the model in shape. The faces will fold better if the Bristol board is cut half way through on the folding edge.

397. For convenient reference the characteristic features of the Third Angle Method, all of which have now been fully illustrated, may thus be briefly summarized:

(a) The various views of the object are so grouped that the *plan* or *top view* comes *above* the front elevation; that of the bottom *below* it; and analogously for the projections of the right and left sides.

(b) Central, reference planes are taken through the various views, and, in each view, the distance of any point from the trace of the central plane of that view is obtained by direct transfer, with the dividers, of the distance between the same point and reference plane, as seen in some other view, usually the plan.

398. *To draw a truncated, pyramidal block, having a rectangular recess in its top;* angle of sides, 60°; lower base a rectangle $3'' \times 2''$, having its longer sides at 30° to the horizontal; total height $1\frac{9}{10}''$; recess $1\frac{9}{10}'' \times \frac{9}{10}''$, and $1''$ deep. (Fig. 240.)

The small oblique projection on the right of the plan shows, pictorially, the figure to be drawn.

PROJECTION OF SOLIDS.—WORKING DRAWINGS.

The plan of the lower base will be the rectangle $a\,b\,d\,e$, $3'' \times 2''$, whose longer edges are inclined $30°$ to the horizontal.

Take AB and mn as the H-traces of auxiliary, vertical planes, perpendicular to the side and end faces of the block. Then the sloping face whose lower edge is $d\,e$, and which is inclined $60°$ to H, will have d_1y for its trace on plane mn. A parallel to mn and $\frac{7}{16}''$ from it will give s_1, the auxiliary projection of the upper edge of the face $s\,r\,e\,d$, whence $s\,r$—at first indefinite in length—is derived, parallel to $d\,e$. Similarly the end face $b\,t\,s\,d$ is obtained by projecting $d\,b$ upon AB at b_1, drawing $b_1 z$ at $60°$ to AB and terminating it at s_2, by CD, drawn at the same height ($\frac{7}{16}''$) as before. A parallel to $b\,d$ through s_2 intersects $r\,s_1$ at s, giving one corner of the plan of the upper base, from which the rectangle $s\,t\,u\,v$ is completed, with sides parallel to those of the lower base.

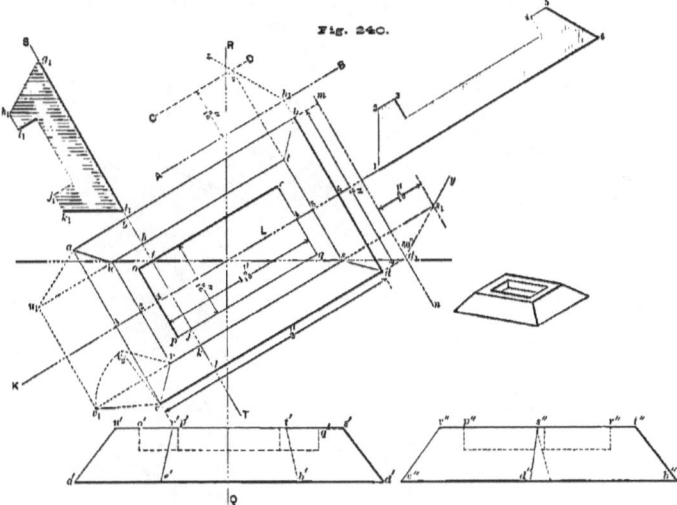

Fig. 240.

As the recess has vertical sides we may draw its plan, $o\,p\,q\,r$, directly from the given dimensions, and show the depth by short-dash lines in each of the elevations.

The ordinary elevations are derived from the plan as in preceding problems; that is, for the front elevation, $a'a's'd'$, by verticals through the plans, terminating according to their height, either on $a'd'$ or on $a's'$, $\frac{7}{16}''$ above it. For the side elevation, $c''v''t''b''$, with the heights as in the front elevation, the distances to the right or left of s'' equal those of the plans of the same points from $s\,i$, regarding the latter as the h. t. of a central, vertical plane, parallel to V.

The plane ST of right section, perpendicular to the axis KL, cuts the block in a section whose true size is shown in the line-tinted figure $g_1 h_1 k_1 l_1$, and whose construction hardly needs detailed treatment after what has preceded. The shaded, longitudinal section, on central, vertical plane KL, also interprets itself by means of the lettering.

142   THEORETICAL AND PRACTICAL GRAPHICS.

The *true size of any face*, as $auve$, may be shown by rabatment about a horizontal edge, as ae. As v is actually $\frac{8}{10}''$ above the level of e, we see that ve (in space) is the hypotenuse of a triangle of base re and altitude $\frac{8}{10}''$. Construct such a triangle, rv_xe, and with its hypotenuse r_xe as a radius, and e as a centre, obtain r_1 on a perpendicular to ae through v and representing the path of rotation. Finding u_1 similarly we have au_1v_1e as the actual size of the face in question.

If more views were needed than are shown the student ought to have no difficulty in their construction, as no new principles would be involved.

399. *To draw a hollow, pentagonal prism,* 2″ *long; edges* to be horizontal and inclined 35° to V; *base,* a regular pentagon of 1″ *sides; one face* of the prism to be inclined 60° to H; distance between inner and outer faces, ¼″.

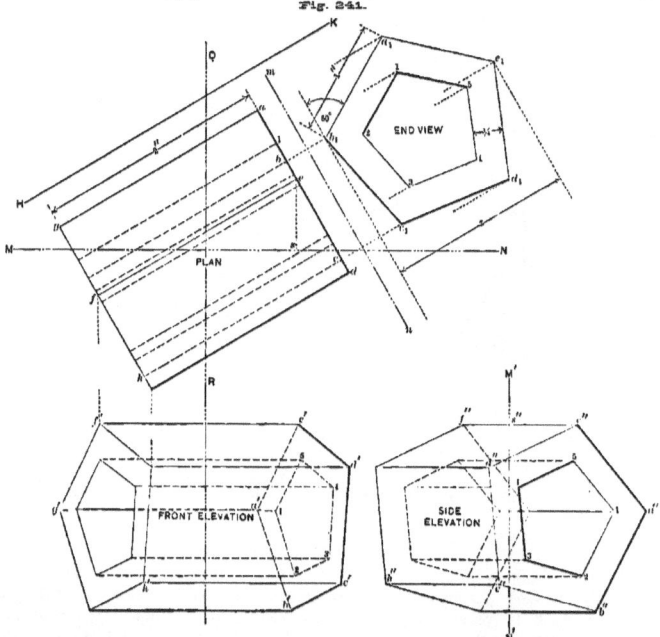

Fig. 241.

In Fig. 241 let HK be parallel to the plans of the axis and edges; it will make 35° with a horizontal line. Perpendicular to HK draw mn as the h. t. of an auxiliary, vertical plane, upon which we may suppose the base of the prism projected. In end view all the *faces* of the prism would be seen as *lines,* and all the *edges* as *points.* Draw a_1b_1, one inch long and at 60° to mn, to represent the face whose inclination is assigned. Completing the inner and outer pentagons, allowing ¼″ for the distance between faces, we have the end view complete. The plan is then obtained by drawing parallels to HK through all the vertices of the end view, and terminating all by vertical planes, ad and gh, parallel to mn and $2''$ apart.

The elevations will be included between horizontal lines whose distance apart is the extreme height z of the end view; and all points of the *front elevation* are on verticals through their plans, and at heights derived from the end view. The most expeditious method of working is to draw a horizontal reference line, like that of Fig. 243, which shall contain the lowest edge of each elevation; measuring upward from this line lay off, on some random, vertical line, the distance of each point of the end view from a line (as the parallel to mn through b_1 in Fig. 241, or xy in Fig. 243) which represents the intersection of the plane of the end view by a horizontal plane containing the lowest point or edge of the object; horizontal lines, through the points of division thus obtained, will contain the projections of the corners of the front elevation, which may then be definitely located by vertical lines let fall from the plans of the same points. For example, c' and f', Fig. 241, are at a height, z, above the lowest line of the elevation, equal to the distance of c, from the dotted line through b_1; or, referring to Fig. 243, which, owing to its greater complexity, has its construction given more in detail, the distance upward from M to line G is equal to y_1y_2 on the end view; from M to Q equals q_1q_2, and similarly for the rest.

Since the profile plane is omitted in Fig. 241 we take $M'N'$ to represent the trace upon it of the auxiliary, central, vertical plane whose h. t. is MN; as already explained, all points of the *side elevation* are then at the same level as in the front elevation, and at distances to the right or left of $M'N'$ equal to the perpendicular distances of their *plans* from MN. For example, $c''s''$ equals cs.

The shade lines are located on the end view on the assumption that the observer is looking toward it in the direction HK.

400. *Projections of a hollow, pentagonal prism, cut by a vertical plane oblique to V.* Letting the data for the prism be the same as in the last problem, we are to find what modification in the appearance of the elevations would result from cutting through the object by a vertical plane PQ (Fig. 242) and removing the part $hxdi$ which lies in front of the plane of section.

Each vertex of the section is on an edge of the elevation and is vertically below the point where PQ cuts the plan of the same edge; the student can, therefore, readily convert the elevations of Fig. 241 into reproductions of those of Fig. 242 by drawing across the plan of Fig. 241 a trace PQ, similarly situated to the PQ of Fig. 242. Supposing that done, refer in what follows to both Figures 241 and 242.

Since PQ contains h we find h' as one corner of the section. Both ends of the prism being vertical, they will be cut by the vertical plane PQ in vertical lines; therefore $h'l'$ is vertical until the top of the prism is reached, at l'. Join l' with x', the latter on the vertical through x—the intersection of PQ with the right-hand top edge cd, $c'd'$. From x the cut is vertical until the interior of the prism is reached, at o', on the line 5-4. We next reach w' on edge No. 4. The line $o'w'$ has to be parallel to $x'l'$ (two parallel planes are cut by a third in parallel lines); but from w' the interior edge of the section is not parallel to $l'h'$, since PQ is not cutting a vertical end, but the inclined, interior surface. The other points hardly need detailed description, being similarly found.

The side elevation is obtained in accordance with the principle fully described in Art. 396 and summarized in Art. 397 (b). $M'N'$ represents the same plane as MN; $c''s''$ equals cs, and analogously for other points.

401. In his elementary work in projections and sections of solids the student is recommended to lay an even tint of burnt sienna, medium tone, over the projections of the object, after which

144   THEORETICAL AND PRACTICAL GRAPHICS.

any *section* may be line-tinted; and, if he desires to further improve the appearance of the views, distinctions may be made between the tones of the various surfaces by overlaying the burnt sienna with flat or graded washes of India ink.

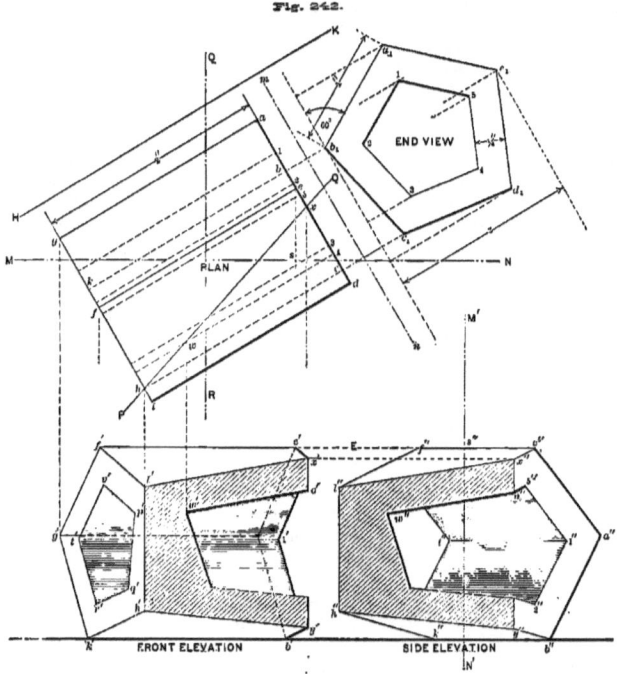

402. *Projections of an L-shaped block, after being cut by a plane oblique to both* V *and* H; *the block also to be inclined to* V *and* H, *and to have running through it two, non-communicating, rectangular openings, whose directions are mutually perpendicular.*

If the dotted lines are taken into account the front elevation in Fig. 243 gives a clear idea of the shape of the original solid. The end view and plan give the dimensions.

Requiring the horizontal edges of the block to be inclined 30° to V, draw the first line xy at 60° to the horizontal; the *plans* of all the horizontal edges will be perpendicular to xy.

Let the inclination of the bottom of the block to H be 20°. This is shown in the end view by drawing $m_1 p_1$ at 20° to xy. All the edges of the end view of the object will then be parallel or perpendicular to $m_1 p_1$, and should be next drawn to the given dimensions.

The central opening, $b_1 d_1 n_1 o_1$, through the larger part of the block, has its faces all $\frac{1}{4}''$ from the outer faces. In the plan this is shown by drawing the lines lettered af at a distance of $\frac{1}{4}''$ from the boundary lines, which last are indicated as $1\frac{1}{4}''$ apart.

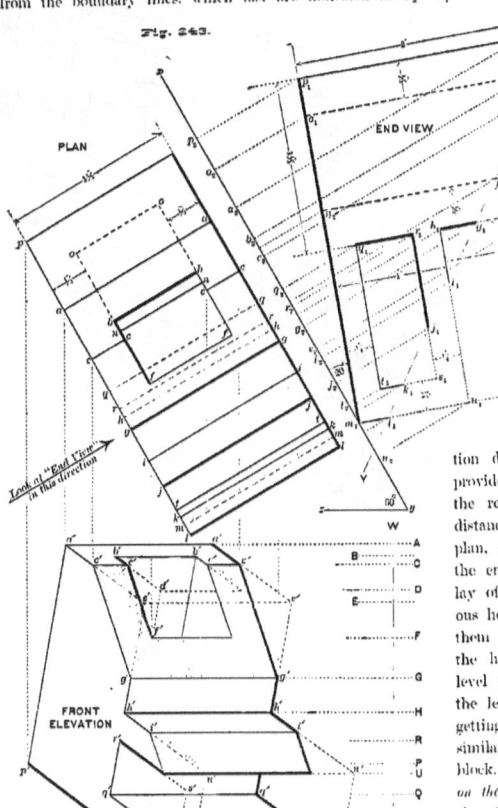

Fig. 243.

The opening $q_1 e_1 s_1 t_1$ has three of its faces $\frac{1}{4}''$ from the outer surfaces of the block, while the fourth, $q_1 e_1$, is in the same plane as the outer face $h'_1 c_1$.

The cutting plane XY gives a section which is seen in end view in the lines $c_1 g_1$, $i_1 j_1$ and $k_1 l_1$; while in plan the section is projected in the shaded portion, obtained, like all other parts of the plan, by perpendiculars to xy from all the points of the end view.

For the front elevation draw first the "reference line." To provide against overlapping of projections the reference line should be at a greater distance below the lowest point, l, of the plan, than the greatest height ($a_1 a_2$) of the end view above xy. Then on MW lay off from M the heights of the various horizontal edges of the block, deriving them from the end view. Thus $a_1 a_2$ is the height of Aa' from M; from M to level B equals $b_1 b_2$, etc. Next project to the level A from points aa of the plan, getting edge $a'a'$ of the elevation, and similarly for all the other corners of the block. Notice that all lines that are parallel on the object will be parallel in each projection (except when their projections coincide); also that in the case of sections, those outlines will be parallel which are the intersection of parallel planes by a third plane.

These principles may be advantageously employed as checks on the accuracy of the construction by points. The construction of the side elevation is left to the student.

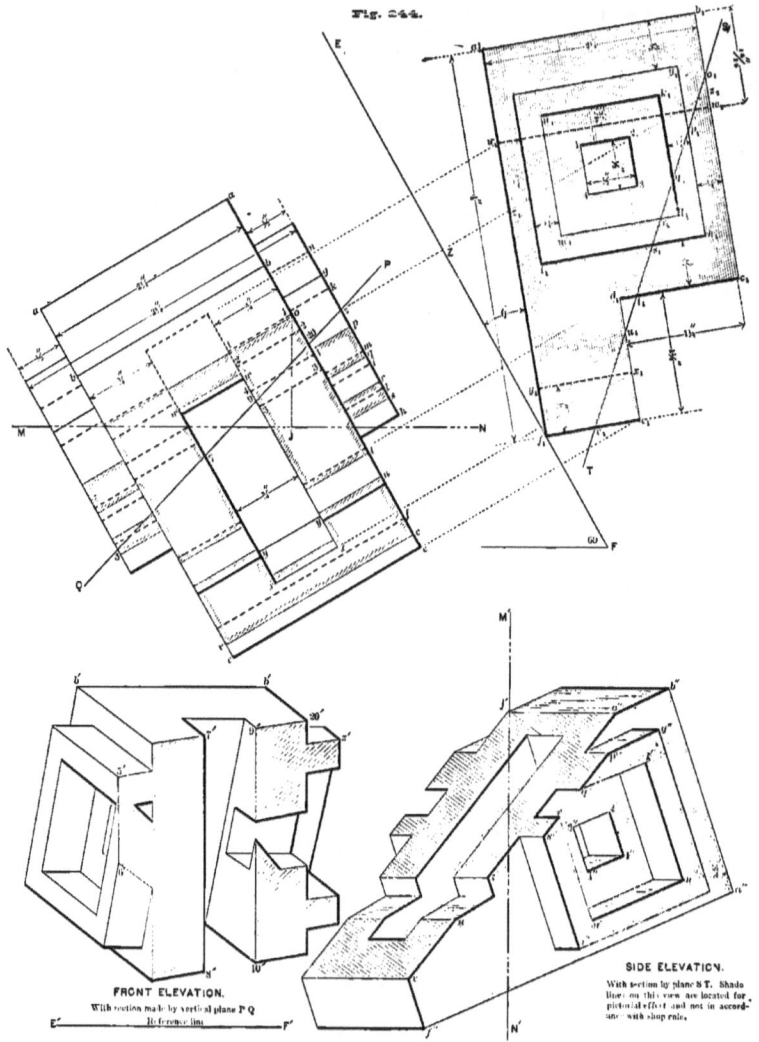

Fig. 244.

FRONT ELEVATION.
With section made by vertical plane P Q
Reference line

SIDE ELEVATION.
With section by plane S T. Shade lines on this view are located for pictorial effect and not in accordance with slope rule.

403. *Projections and sections of a block of irregular form, with two mutually perpendicular openings through it, and with equal, square frames projecting from each side.*

In Fig. 244 the side elevation shows clearly the object dealt with, while we look to the end view for most of the dimensions. The large central opening extends from $w_1 w_2$ to $x_1 y_1$. The width of the main portion of the block is shown in plan as $2\frac{1}{4}''$, between the lines lettered $a e$. The square frames project $\frac{1}{4}''$ from the sides, while the width of the central opening between the lines $w x$ is $\frac{3}{4}''$.

Two section planes are indicated, $S T$ across the end view, and $P Q$—a vertical plane—across the plan; the section made by plane $S T$ is, however, shown only in fringed outline on the plan, though fully represented on the side elevation. The front elevation shows the section made by plane $P Q$, with the visible portion of that part of the object that is behind the cutting plane.

Although detailed explanation of this problem is unnecessary after what has preceded, yet a brief recapitulation of the various steps in the construction of the views may be appreciated by some, before passing on to a more advanced topic.

(a) $E F$, the first line to draw, is the trace of the vertical plane on which the end view is projected, and is at an angle of 60° to a horizontal line in order that the edges of the object (as $a a$, $b b \cdots e e$) may be inclined at 30° to the front vertical plane, which we may assume as one of the conditions of the problem.

(b) A rotation of the object through an angle $\theta°$ about a horizontal axis that is perpendicular to $E F$, as, for example, the edge through f, is shown by the inclination of the end view to $E F$ at an angle $a_1 f_1 E = \theta°$.

(c) Drawing the end view at the required angle to $E F$ we next derive the plan therefrom by perpendiculars to $E F$, terminating them on parallels to $E F$ (as the lines $a e$, $w x$, $a b$, etc.,) whose distances apart conform to given data.

(d) *The elevations.* For these a common reference line $E' F' f''$ is taken, horizontal, and sufficiently below the plan to avoid an overlapping of views.

For the *front elevation* any point, as b', is found vertically below its plan b, and is as far from $E' F''$ as b_1 is—perpendicularly—from $E F$.

The height at which the section plane $P Q$ cuts any line is similarly obtained. Thus at z it cuts the vertical end face of the block in a line which is carried over on the end view in the indefinite line $Z z_1$; the portions of $Z z_1$ which lie on the end view of the *frame* $g_1 h_1 i_1$ are the only real parts to transfer to the front elevation, and are seen on the latter, vertically below z and running from z' down; their distances from $E' F''$ being simply those from Z, on the end view, transferred.

The side elevation. Any point or edge is at the same *level* on the side elevation as on the front; hence the edge through b'' is on $b' b'$ produced. The distances to the right or left of $M' N'$ equal those of the corresponding points on the plan from $M N$; thus $a'' j'$ equals $o j$, etc.

404. *Changed planes of projections.* In the problems of Arts. 339–403 the employment of an "end view"—which was simply an *auxiliary elevation*—has prepared the student for the further use of planes other than the usual planes of projection; and if the *auxiliary plan* is now mastered he is prepared to deal with any case of rotation of object about vertical or horizontal axes, since new and properly located planes of projection are their practical equivalent.

In Fig. 245 the object is represented in its initial position by the line-tinted figures marked "first plan" and "first elevation." The third and fourth elevations show somewhat more pictorially that it is a hollow, truncated, triangular prism, having through it a rectangular opening that is perpendicular to the front and rear faces.

148 THEORETICAL AND PRACTICAL GRAPHICS.

Fig. 245.

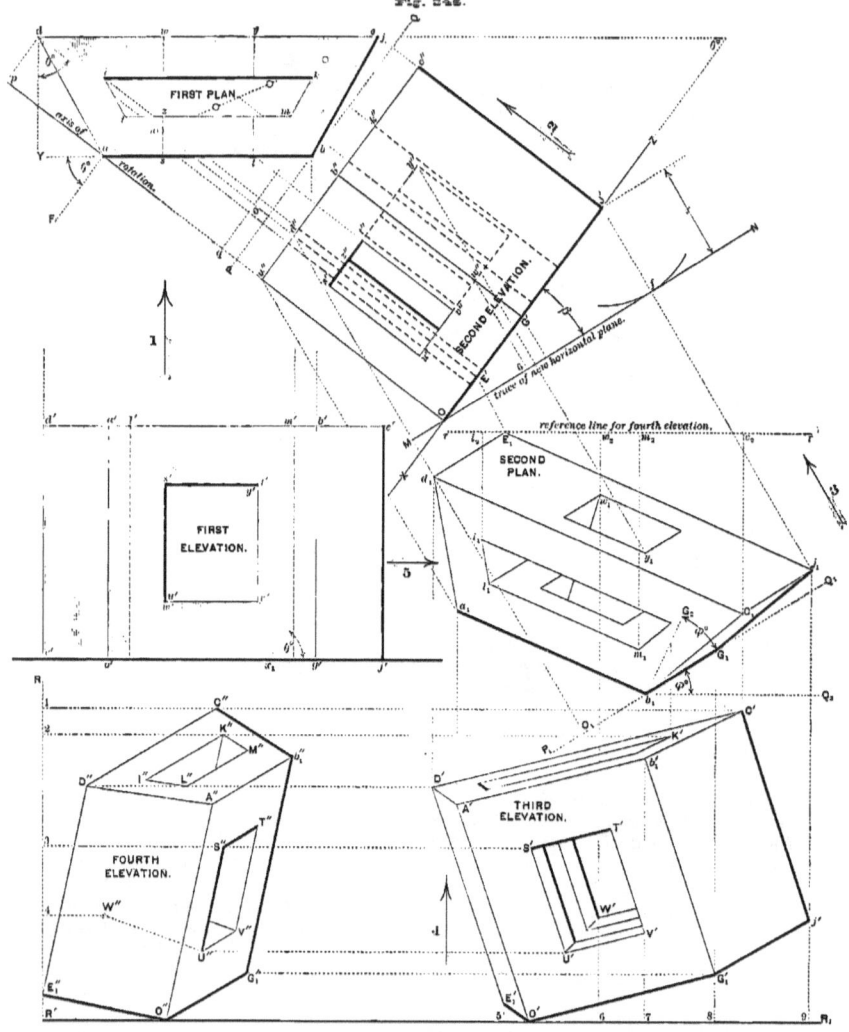

CHANGING PROJECTION-PLANE EQUIVALENT TO ROTATING OBJECT. 149

(a). *Rotation about a vertical axis*, or its equivalent, *a change in the vertical plane of projection*. Result: *second elevation* derived from *first plan and elevation*.

Let the axis be one of the vertical edges of the object, as that at d in the first plan; also let the rotation be through an angle $Yd o$ or $\theta°$, (θ being taken, for convenience, equal to the angle YaF, which—with the line pq—will be employed in a later construction). If we were actually to rotate the object through an angle θ the new *plan* would be the exact counterpart of the first, but its horizontal edges would make an angle θ with their former direction, and the new *elevation* would partly overlap the first one. To avoid the latter unnecessary complication, as also the duplication of the plan, we make the first plan do double duty, since we can accomplish the equivalent of rotation of the object by taking a *new vertical plane* that makes an angle θ with the plane on which the first elevation was made. This equivalence will be more evident, if some small object, as a piece of india rubber, is placed on the "first plan" with its longer edges parallel to ah, and is then viewed in the direction of arrow No. 2 through a pane of glass standing vertically on XZ; after which turn both the object and the glass through the angle θ until the glass stands vertically on $e'j'$ and then view in the direction of arrow No. 1.

The second plane may be located *anywhere*, as long as the angle θ is preserved; XZ, making angle θ at x_1 with $e'j'$, is, therefore, a *random position* of the new plane, and the projection upon it is our "second elevation."

Since the *heights* of the various corners of an object remain unchanged during rotation about a vertical axis we will find all points of the second elevation at distances from the reference line XZ that are derived from the first elevation, and laid off on lines drawn perpendicular to XZ from the vertices of the plan; thus aO is perpendicular to XZ, and Oa'' equals $a'a'$; $e''J' = e'j'$, etc.

(b). *Rotation about a horizontal axis*, or its equivalent, *the adoption of a new horizontal plane*. Result: *second plan* derived from *first plan and second elevation*.

Having in the last case illustrated the method of complying with the condition that rotation should occur *through a given angle* (which is incidentally shown again, however, in the next construction) we now choose an axis pq so as to illustrate a different kind of requirement, viz.: that during rotation the *heights* of any two points of the object, which were at first *at the same level as the axis*, shall be in some predetermined ratio, regardless of the *amount* of rotation. In the figure it is assumed that $e'(d)$ is to be at one-fifth the height of $j'j$, and that rotation shall occur about an axis passing through the lower end a' of the vertical edge at a. By drawing ad and aj, dividing the latter into five equal parts, and joining d with u—the first point of division from a—we obtain the direction du, parallel to which the axis pq is drawn through a. The distance dp is then one-fifth of jq, and they shorten in the same ratio, as rotation occurs.

After locating the axis the next step is, invariably, the drawing of an elevation upon a plane perpendicular to the axis. This we happen, however, to have already in our "second elevation," having, in the interest of compactness, so taken θ in the preceding case that the vertical plane XZ would be perpendicular to the axis we are now ready to use.

Any rotation of the object about pq will, evidently, not change the *form* of the "second elevation" but simply incline it to XZ. But, as before, instead of actually rotating the object, which would probably give projections overlapping those from which we are working, we adopt a new plane MN as a horizontal plane of projection, so taken that it fulfills either of the following conditions: (a) that the object should be rotated about pq through an angle β; (b) that the corner J' should be higher than O by an amount x, MN being drawn tangent to an arc having J' for its centre, and $J'j$ (equal to x) for its radius.

Reference to Fig. 246 may make it clearer to some that MN is the trace of the new plane upon the vertical plane whose h. t. is XZ; that ON lies vertically below the line XZ and is as truly perpendicular to the axis of rotation as is NZ; also that in Fig. 245 a view in the direction of arrow No. 3 (i. e., perpendicular to MN) is equivalent to a view perpendicular to the plane V in Fig. 246 after the whole assemblage of planes and object has been rotated together about HOH until the "new plane" takes the position out of which the first horizontal plane has just been rotated.

Fig. 246.

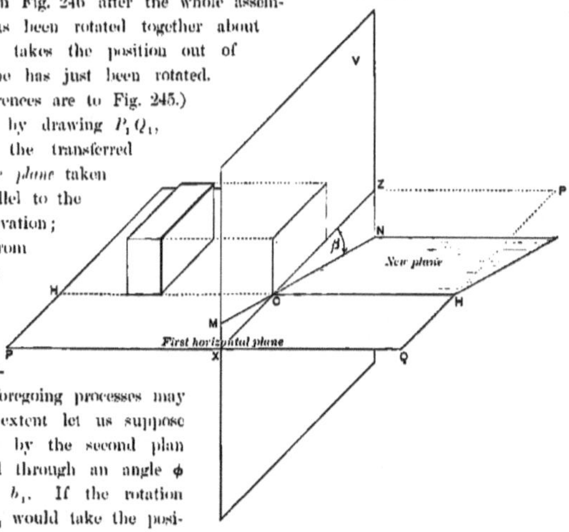

(The remainder of the references are to Fig. 245.)

The *second plan* is obtained by drawing $P_1 Q_1$, parallel to MN, to represent the transferred trace PQ of a vertical *reference plane* taken through some edge b and parallel to the plane ZON of the second elevation; then any point d_1 is as far from $P_1 Q_1$ as the same point d, on the first plan is from PQ; that is, $d_1 O_1$ equals od, and similarly for other points.

(c) *Further rotation about vertical axes*. To show how the foregoing processes may be duplicated to any desired extent let us suppose that the object, as represented by the second plan and elevation, is to be rotated through an angle ϕ about a vertical line through b_1. If the rotation actually occurred, the plan $b_1 G_1$ would take the position $b_3 G_3$, and the other lines of the plan would take corresponding positions in relation to a vertical plane on $P_1 Q_1$. A new vertical plane on $b_1 Q_3$, at an angle ϕ to $b_1 G_1$, will, however, evidently hold the same relation to the plan as it stands, and transferring such new plane forward to $O'R_1$ we then obtain the points of the new (third) elevation by letting fall perpendiculars to $O'R_1$ from the vertices of the second plan, and on them laying off heights above $O'R_1$ equal to those of the same points above MN in the *second elevation*. Thus $j'9$ equals $J'f$; $H'''6$ in the fourth equals $H'''6$ in the second.

The *fourth elevation* is a view in the direction of arrow No. 5, giving the equivalent of a ninety-degree rotation of the object from its last position. To obtain it take a reference line rr through some point of the second plan, and parallel to $O'R_1$; then RR' represents the vertical plane on rr, transferred. From RR' lay off—on the *levels* of the same points in the third elevation—distance $1C''' = c_3C_1$; $4H''' = W_2 W_1$, as in preceding analogous constructions.

THE DEVELOPMENT OF SURFACES.

405. *The development of surfaces* is a topic not altogether new to the student who has read Chapter V and the earlier articles of this chapter;[*] so far, however, it has occurred only incidentally, but its importance necessitates the following more formal treatment, which naturally precedes problems on the *interpenetration of surfaces*, of which a development is usually the practical outcome.

---

[*] The following articles should be carefully reviewed at this point: 139; 191; 344–6; 389, and Case 6 of Art. 386.

THE DEVELOPMENT OF SURFACES.

A *development* of a surface, using the term in a practical sense, is a piece of cardboard or, more generally, of sheet-metal, of such shape that it can be either directly *rolled up* or *folded* into a model of the surface. Mathematically, it would be the *contact-area*, were the surface rolled out or unfolded upon a plane.

The "shop" terms for a developed surface are "surface *in the flat*," "stretch-out," "roll-out"; also, among sheet-metal workers it is called a *pattern*; but as *pattern-making* is so generally understood to relate to the patterns for castings in a foundry, it is best to employ the qualifying words *sheet-metal* when desiring to avoid any possible ambiguity.

406. The mathematical nature of the surfaces that are capable of development has been already discussed in Arts. 344–346. Those most frequently occurring in engineering and architectural work are the right and oblique forms of the pyramid, prism, cone and cylinder.

407. In Art. 120 the *development of a right cylinder* is shown to be a *rectangle* of base equal to $2 \pi r$ and altitude h, where h is the height of the cylinder and r is the radius of its base.

408. The *development of a right cone* is proved, by Art. 191, to be a *circular sector*, of radius equal to the slant height R of the cone, and whose angle θ is found by means of the proportion $R : r :: 360° : \theta$; r being the radius of the base of the cone.

409. The *development of a right pyramid* is illustrated in Art. 389 and in Case 6 of Art. 396.

410. We next take up right and oblique prisms, and the oblique pyramid, cone and cylinder; while for the sake of completeness, and departing in some degree from what was the plan of this work when Art. 345 passed through the press, the *regular solids* will receive further treatment, and also the *developable helicoid*.

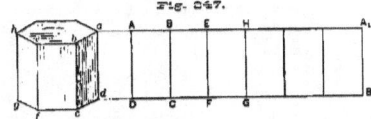

Fig. 247.

411. The *development of a right prism*. Fig. 247, represents a regular, hexagonal prism. The six faces being equal, and $e b c f$ showing their actual size, we make the rectangles $ABCD$, $BEFC$, etc., each equal to $e b c f$; then AA_1 equals the perimeter of the upper base, and we have the rectangle AA_1B_1D for the development sought.

412. The *development of a right prism below a cutting plane*. Taking the same prism as in the last article develop first as if there were no section to be taken into account. This gives, as before, a rectangle of length AA_1 and of altitude $a d$, divided into six equal parts. Then project, from each point where the plane cuts an edge, to the same edge as seen on the development.

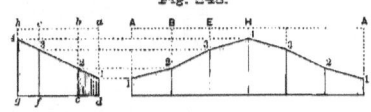

Fig. 248.

413. *Right Section. Rectified Curve. Developed Curve.* A plane perpendicular to the axis of a surface cuts the latter in a *right section*. The bases of right cones, pyramids, cylinders and prisms fulfil this condition and require no special construction for their determination; but the development of an oblique form usually involves the construction of a right section and then the laying off on a straight line of a length equal to the perimeter of such section. Should the right section be a *curve* its equivalent length on a straight line is called its *rectification*, which should not be confounded with its *development*, the latter not being necessarily straight.

414. The *development of an oblique prism, when the faces are equal in width*. In Fig. 249, an oblique, hexagonal prism is shown, with x for the width of its faces. Since the perimeter of a right section would evidently equal $6x$ we may directly lay off x six times on some perpendicular

to the edges, as that through a. The seven parallels to ab, drawn at distances x apart, will contain the various edges of the prism as it is rolled out on the plane; and the position of the extremities are found by perpendiculars from their original positions. The initial position $a_1 b_1$ is *parallel to* but at any distance from ab.

415. *The development of an oblique prism whose faces are unequal in width.*

In Fig. 250 $c'd'h'g'$ is the elevation of the prism; ap a plane of right section. To get the true shape, 1-2-3-4, of the right section, we require $abhfc$ — plan of the prism.* Assuming that to have been given, imagine next a vertical reference-plan standing on ab. The right section plane np cuts the edge $c'd'$ at a, which is at a distance x in front of the assumed reference-plane. Make $n2 = x$. Similarly make $n3 = y$, and $p4 = z$; then 1-2-3-4 is the right section, seen in its true size after being revolved about the trace of the right-section plane upon the assumed reference-plane.

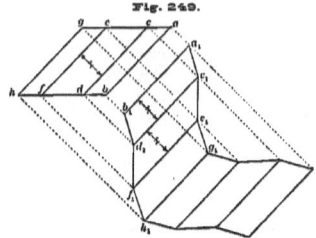

Fig. 249.

Fig. 250.

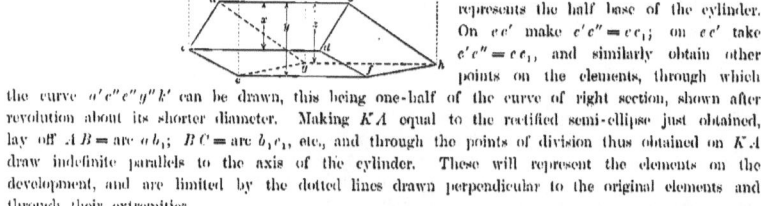

Prolong pn indefinitely, and on its extension make $1'-2' = 1-2$; $2'3' = 2-3$, etc. Parallels to $c'd'$ through the points of division thus obtained will contain the edges of the developed prism, and their lengths are definitely determined by perpendiculars, as $h'h''$, $f'f''$, from the extremities of the original edges.

416. *The development of an oblique cylinder.* Let $am'n'k$, Fig. 251, be an oblique cylinder with circular base. Take any plane of right section, as $n'k'$. Draw various elements, as those through b', c', etc., and from their lower extremities erect perpendiculars to ak, as ce_1, terminating them on the arc af_1k, which represents the half base of the cylinder. On ee' make $c'e'' = ee_1$; on ee' take $c'c'' = ee_1$, and similarly obtain other points on the elements, through which the curve $a'c''e''g''k'$ can be drawn, this being one-half of the curve of right section, shown after revolution about its shorter diameter. Making KA equal to the rectified semi-ellipse just obtained, lay off AB = arc ab_1; BC = arc $b_1 c_1$, etc., and through the points of division thus obtained on KA draw indefinite parallels to the axis of the cylinder. These will represent the elements on the development, and are limited by the dotted lines drawn perpendicular to the original elements and through their extremities.

The area $a_1 k_2 NM$ is the development of one-half of the cylinder, the shaded area representing all between $a'k'$ and the base ak.

* In the interest of compactness the "First Angle" position of the views is employed in Figs. 250, 253 and 255.

THE DEVELOPMENT OF SURFACES.

417. *The development of an oblique pyramid.* The development will evidently consist of a series of triangles having a common vertex. To ascertain the length of any edge we may carry it *into* or *parallel to* a plane of projection. Thus in Fig. 252 the edge vb is carried into the vertical plane at $v b''$. Its true length is evidently the hypothenuse of a right-angled triangle of base $ob = ab''$, and altitude vo.

In Fig. 253 a pyramid is shown in plan and elevation. Making $a a'' = v a$ we have $v' a''$ for the actual length of edge $v' a'$, a construction in strict analogy to that of Fig. 252. The plan vb being parallel to the base line shows that $v' b'$ is the actual length of that edge. By carrying

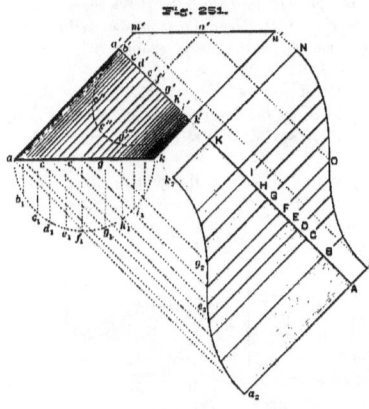

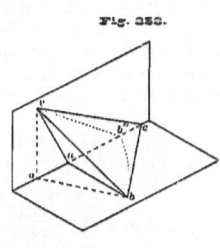

vc to vc_1, where it becomes parallel to V, and then projecting c_1 to c'' we get $v'c''$ for the true length of edge $v'c'$.

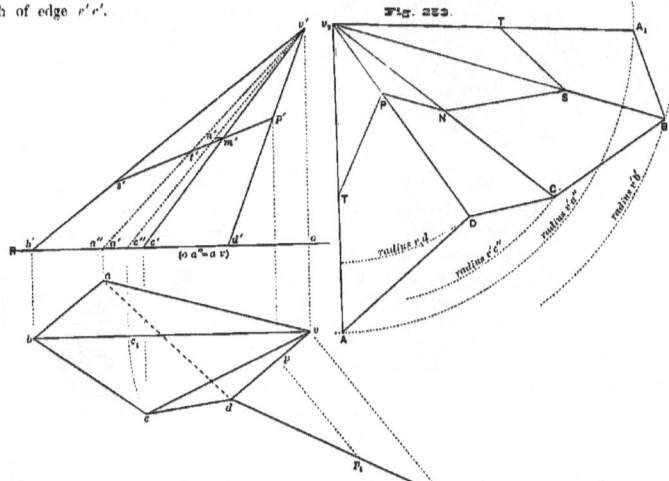

To illustrate another method make $v c_1 = v' o$; then $v_1 d$ ⤳ is the real length of $v' d'$, shown by rabatment into H.

154   *THEORETICAL AND PRACTICAL GRAPHICS.*

For the development take some point c_2 and from it as a centre draw arcs having for radii the ascertained lengths of the edges. Letting $c_2 A$ represent the initial edge of the development take A as a centre, ad as a radius, and cut the arc of radius $c_1 d$ at D; then $A c_2 D$ is the *development* of the face $a c d$, $a'c'd'$. With centre D and radius dc obtain C on the arc of radius $c'c''$; similarly for the remaining faces, completing the development $c_2 - A D \ldots A_1$.

The shaded area $c_2 - T P \ldots T$ is the development of that part of the pyramid above the oblique plane $s'p'$, found by laying off, on the various edges as seen in the development, the distances along those edges from the vertex to the cutting plane; thus $c_2 S = c's'$; $c_2 N = c'n'$, the real length of $r'a'$; $c_2 P = c_1 p_1$, the length of $r'p'$, etc.

418. *The development of an oblique cone.* The usual method of solving this problem gives a result which, although not mathematically exact, is a sufficiently close approximation for all practical purposes. In it the cone is treated as if it were a pyramid of many sides. The length of any element is then found as in the last problem. Thus in Fig. 254 an element rc is carried to rc'' about the vertical axis ro.

In Fig. 255 we have $r'a g$ for the elevation of the cone, and $a - abc \ldots g$ for the half plan. Make $ob'' = ob$; then $r'b''$ is the real length of the element whose plan is ab. Similarly, c, d, e and f are carried by arcs to ag and there joined with r'.

For the development make $r_1 A$ equal and parallel to $r'a$, and at any distance from it. With r_1 as a centre draw arcs with radii equal to the true lengths of the elements; then, as in the pyramid, make $AB = $ arc ab; $BC = $ arc bc, etc.

The greater the number of divisions on the semi-circle $ab \ldots g$ the more closely will the development approximate to theoretical exactness.

419. *The five regular convex solids*, with the forms of their developments, are illustrated in Figs. 256-265. They have already been defined in Art. 345, and that five is their limit as to number is thus shown: The faces are to be equal, regular polygons, and the sum of the plane angles forming a solid angle must be less than four right angles; now as the angles of *equilateral triangles* are 60° we may evidently have groups of three, four or five and not exceed the limit; with *squares* there can be groups of three only, each 90°; with regular *pentagons*, their interior angles being 108°, groups of three; while hexagons are evidently impracticable, since three of their interior angles would exactly equal four right angles, adapting them perfectly—and only—to plane surfaces. (See Fig. 131.)

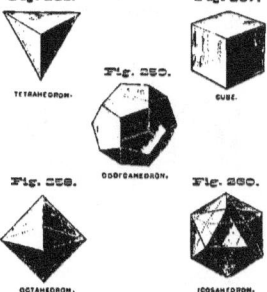

THE FIVE REGULAR CONVEX SOLIDS.

The dihedral angles between the adjacent faces of regular solids are as follows: 70° 31′ 44″ for the tetrahedron; 90° for the cube; 109° 28′ 16″ for the octahedron; 116° 33′ 54″ for the dodecahedron; and 138° 11′ 23″ for the icosahedron.

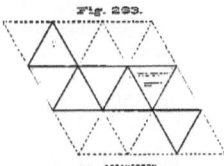

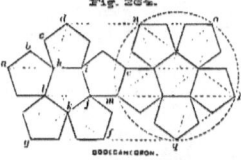

A sphere can be inscribed in each regular solid and can also as readily be circumscribed about it.

The relation between d, the diameter of a sphere, and e, the edge of an inscribed regular solid, is illustrated graphically by Fig. 266, but may be otherwise expressed as follows:

For the *tetrahedron* $d : e :: \sqrt{3} : \sqrt{2}$; for the *cube* $d : e :: \sqrt{3} : 1$
"   " *octahedron* $d : e :: \sqrt{2} : 1$;   "   " *dodecahedron* e = the greater segment of the edge of an inscribed cube when the latter has been medially divided, that is, in extreme and mean ratio.

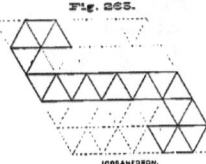

For the *icosahedron* e = the chord of the arc whose tangent is d; i. e., the chord of 63° 26′ 6″.

Reference to Figs. 256–260 and the use of a set of cardboard models which can readily be made by means of Figs. 261–265 will enable the student to verify the following statements as to those ordinary views whose construction would naturally precede the solution of problems relating to these surfaces.

In all but the tetrahedron each face has an equal, opposite, parallel face, and except in the cube such faces have their angular points alternating. (See Figs. 266, 267, 268.)

The *tetrahedron* projects as in Fig. 256, upon a plane that is parallel to either face.

The *cube* projects in a square upon a plane parallel to a face, while on a plane perpendicular to a body diagonal it projects as a regular hexagon, with lines joining three alternate vertices with the centre.

The *octahedron*, which is practically two equal square pyramids with a common base, projects in a square and its diagonals, upon a plane perpendicular to either body diagonal; in a rhombus and shorter diagonal when the plane is parallel to one body diagonal and at 45° with the other

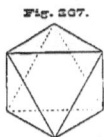

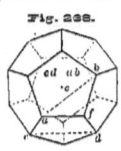

two; and (as in Fig. 267) in a regular hexagon with inscribed triangles (one dotted), when it is projected upon a plane parallel to a face.

The *dodecahedron* projects as in Fig. 268 whenever the plane of projection is parallel to a face.

Fig. 269 represents the *icosahedron* projected on a plane parallel to a face, and Fig. 269 when the projection-plane is perpendicular to an axis.

420. *The Developable Helicoid.* When the word *helicoid* is used without qualification it is understood to indicate one of the *warped* helicoids, such as is met with, for example, in screws, spiral staircases and screw propellers. There is, however, a *developable* helicoid, and to avoid confusing it with the others its characteristic property is always found in its name. As stated in Art. 346 it is generated by moving a straight line tangentially on the ordinary helix, which curve (Art. 120) cuts all the elements of a right cylinder at the same angle. Fig. 269 illustrates the completed surface pictorially; Fig. 270 shows one orthographic projection; and in Fig. 271 it is seen in process of generation by the hypothenuse of a right-angled triangle that rolls tangentially on a cylinder.

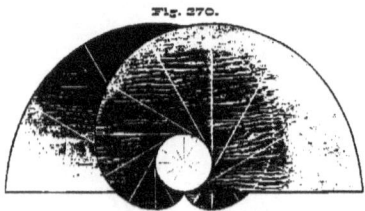

Fig. 270.

The construction just mentioned is based on the property of non-plane curves that at any point the curve and its tangent make the same angle with a given plane; if, therefore, the helix, beginning at *a*, crosses each element of the cylinder at an angle equal to *a b p* in the rolling triangle, the hypothenuse of the latter will evidently move not only tangent to the cylinder, but also to the helix.

The following important properties are also illustrated by Fig. 271:

(a) The involute* of a helix and of its horizontal projection are identical, since the point *b* is the extremity of both the rolling lines, *a b* and *p b*.

(b) The length of any tangent, as *m b*, is that of the helical arc *m a* on which it has rolled.

(c) The horizontal projection *b q* of any tangent *b m* equals the rectification of an arc *a q* which is the *projection* of the helical arc from the initial point *a* to the point of tangency *m*.

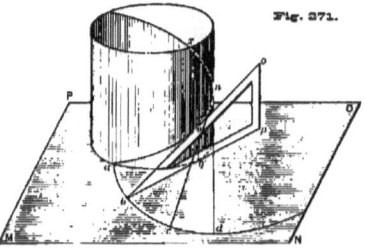

Fig. 271.

The *development of one nappe of a helicoid* is shown in Fig. 273. It is merely the area between a circle and its involute; but the radius ρ, of the base circle, equals $r \sec^2 \theta$,† in which r is

---

* For full treatment of the involute of a circle refer to Arts. 186 and 187.

† This relation is due to considerations of curvature. At any point of any curve its *curvature* is its rate of departure from its tangent at that point. Its *radius of curvature* is that of the *osculatory circle* at that point. (Art. 386.) Now from the nature of the two *uniform* motions imposed upon a point that generates a helix (Art. 120) the curvature of the latter must be *uniform*; and if developed upon a plane *by means of its curvature* it must become a circle—the only *plane* curve of *uniform curvature*. The radius of the developed helix will, obviously, be the radius of curvature of the space helix. Following Warren's method of proof in establishing its value let *a*, *b* and *c* (Fig. 272) be three equi-distant points on a helix, with *b* on the foremost element; then *a'c'* is the elevation of the circle containing these points. One diameter of the circle *a'b'c'* is projected at *b'*. It is the hypothenuse of a right-angled triangle having the chord *bc*, *b'c'*, for its base. Let 2ρ be the diameter of the circle *a'b'c'*; $2r = bd$, that of the cylinder. Using capitals for points *in space* we have $BC^2 = 2\rho \times bd$; also $bc^2 = 2r \times bc$; whence, dividing like members and substituting trigonometric functions (see note p. 31), we have $\rho = r \sec^2 \beta$, in which β is the angle between the line BC and its projection.

Let θ be the inclination of the tangent to the helix at *b'*. If, now, both *A* and *C* approach *B*, the angle β will approach θ as its limit; and when *A*, *B* and *C* become *consecutive* points we will have $\rho = r \sec^2 \theta =$ the radius of the osculatory circle = *the radius of curvature*.

For another proof, involving the radius of curvature of an ellipse, see Olivier, *Cours de Géométrie Descriptive*, Third Ed., p. 197.

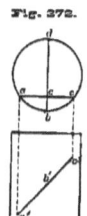

Fig. 272.

THE INTERSECTION OF SURFACES.

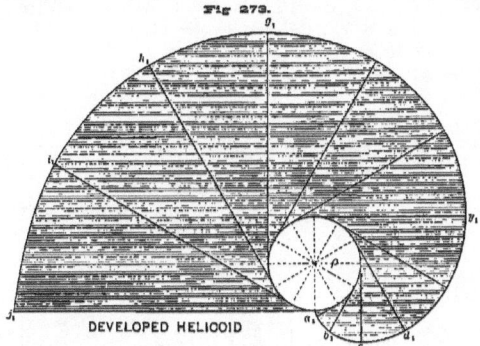

Fig. 273.

DEVELOPED HELICOID

the radius of the cylinder on which the helix originally lay, and θ is the angle at which the helix crosses the elements. To determine ρ draw on an elevation of the cylinder, as in Fig. 274, a line $a\,b_1$ tangent to the helix at its foremost point, as in that position its inclination θ is seen in actual size; then from o, where $a\,b$ crosses the extreme element, draw an indefinite line, $o\,s$, parallel to $c\,d$, and cut it at m by a line $o\,m$ that is perpendicular to $a\,b$ at its intersection with the front element $e f$ of the cylinder; then $o\,m = \rho = r \sec^2 \theta$. For we have

Fig. 274.

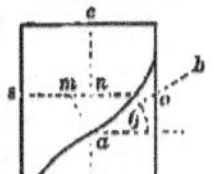

$o a = o n \sec \theta = r \sec \theta$; and $o n (= r) : o a :: o a : o m$; whence $o m = r \sec^2 \theta = \rho$.

The circumference of circle ρ equals $2 \pi r \sec \theta$, the actual length of the helix, as may be seen by developing the cylinder on which the latter lies. The elements which were tangent to the helix maintain the same relation to the developed helix, and appear *in their true length* on the development.

The student can make a model of one nappe of this surface by wrapping a sheet of Bristol board, shaped like Fig. 273, upon a cylinder of radius r in the equation $r \sec^2 \theta = \rho$; or a two-nappcd helicoid by superposing two equal circular rings of paper, binding them on their inner edges with gummed paper, making one radial cut through both rings, and then twisting the inner edge into a helix.

THE INTERSECTION OF SURFACES.

421. *When plane-sided surfaces intersect,* their outline of interpenetration is necessarily composed of straight lines; but these not being, in general, in one plane, form what is called a *twisted* or *warped polygon;* also called a *gauche polygon.*

422. If either of two intersecting surfaces is *curved* their common line will also be curved, except under special conditions.

423. When one of the surfaces is of uniform cross section—as a cylinder or a prism—its end view will show whether the surfaces intersect in a continuous line or in two separate ones. In Cases a, b, c, d and g of Fig. 275, where the end view of one surface either cuts but one limiting line of the other surface or is tangent to one or both of the outlines, the intersection will be a *continuous line.* Two separate curves of intersection will occur in the other possible cases, illustrated by e and f, in which the end view of one surface either crosses both the outlines of the other or else lies wholly between them.

A cylinder will intersect a cone or another cylinder in a *plane curve* if its end view is tangent to the outlines of the other surface, as in d and g, Fig. 275. Two cones may also intersect in a plane curve, but as the conditions to be met are not as readily illustrated they will be treated in a special problem.

Fig. 275.

424. In general, the line of intersection of two surfaces is obtained, as stated in Art. 379, by passing one or more auxiliary surfaces, usually *planes*, in such manner as to cut some easily constructed sections—as straight lines or circles—from each of the given surfaces; the meeting-points of the sections lying in any auxiliary surface will lie on the line sought.

The application of the principle just stated is much simplified whenever any face of one or other of the surfaces is so situated that it is projected in a *line*. This case is amply illustrated in the problems most immediately following.

Fig. 276.

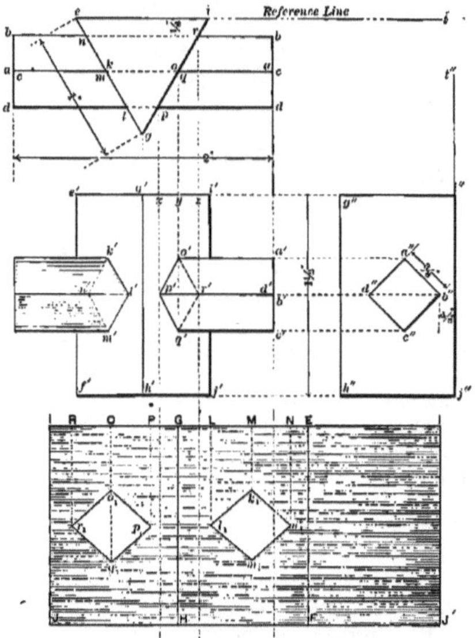

The beginner will save much time if he will letter each projection of a point as soon as it is determined.

425. *The intersection of a vertical triangular prism by a horizontal square prism; also the developments.*

The vertical prism to be $1\frac{1}{2}''$ high and to have one face parallel to V; bases equilateral triangles of 1" side.

The horizontal prism to be 2" long, its basal edges $\frac{3}{8}''$, and its faces inclined 45° to H; its rear edges to be parallel to and $\frac{1}{8}''$ from the rear face of the horizontal prism.

The elevations of the axes to bisect each other.

Draw ei horizontal and 1" long for the plan of the rear face of the vertical prism. Complete the triangle egi and project to levels $1\frac{1}{2}''$ apart, obtaining $e'f'$, $g'h'$, $i'j'$, on the elevation.

Construct an end view $g''i''j''h''$, using $t''j''$ to represent the reference line et, transferred.

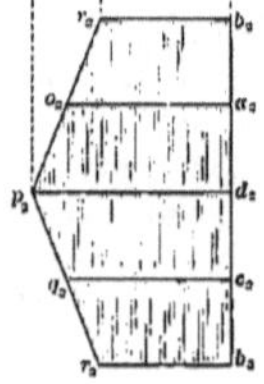

The end view of the horizontal prism is the square $a''b''c''d''$, having its diagonal horizontal and midway between the upper and lower bases of the other prism, and with its corner b'' one-eighth inch from $i''j''$. The plan and front elevation of the horizontal prism are next derived from the end view as in preceding constructions.

Since the lines eg and gi are the plans of vertical faces of one prism we note their intersection by the edges a, b, c and d of the horizontal prism—as at n, m, l, p, q, r—and project to the elevations of the same edges. Thus the edge aa meets the other prism at o and k, which project to the level of a'' at o' and k'. Similarly for the remaining points.

The development of the vertical prism is shown in the shaded rectangle IJ, of length $3gi$ and altitude $e'f'$. (See Art. 411). The openings $o_1p_1q_1r_1$ (and $k_1l_1m_1n_1$ are thus found: For p_1, which represents p', make $Pp_1 = xp'$, and $GP = gp$, the true distance of p' from $g'h'$; similarly, $q_1O = q'y$, and $OG = og$.

THE INTERSECTION OF SURFACES.

The right half of the horizontal prism, $a'a'c'q'$, is developed at $c_2 b_2 b_3 c_3$ after the method of Art. 412.

426. *The intersection of two prisms, one vertical, the other horizontal, each having an edge exterior to the other.* The condition just made will, as already stated (Art. 423), make the result a single warped *polygon*.

Fig. 277. Let $a\, b\, c\, d$, $1'' \times \frac{1}{2}''$, be the plan of the vertical prism, which stands with its broader faces at some convenient angle $e\, w\, g$ to V. From it construct the front and side elevations, taking a reference plane through d for the latter.

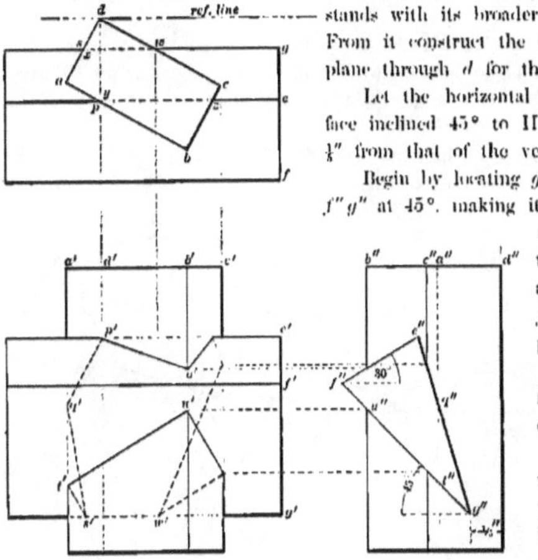

Let the horizontal prism be triangular (isosceles section) one face inclined 45° to H; another 30° to H; the rear edge to be $\frac{1}{4}''$ from that of the vertical prism.

Begin by locating g'' one-fifth inch from the right edge, draw $f''g''$ at 45°, making it of sufficient length to have f'' exterior to the other prism; then $f''e''$ at 30° to H, terminated at e'' by an arc of centre g'' and radius $g''f''$; finally $e''g''$. The edges e', f' and g' of the front elevation are then projected from e'', f'' and g''.

The rear edge g in the plan meets the face ad at s, which projects to s' on the edge g'.

Moving forward from s, the next edge reached, of either solid, is a, of the vertical prism. To ascertain the height at which it meets the other prism we look to the end view, finding q'' for the entrance and t'' for the exit. Being on the way up from g'' to e'' we use q'', reserving t'' until we deal with the face $g''f''$. Projecting q'' over to q' draw $q's'$, dotted, since it is on a rear face.

Returning to a and moving toward b we next reach the edge e, whose intersection p with $a\, b$ is then projected to edge e' at p' and joined with q'.

Fig. 278. For the next edge, b, we obtain o' from the side elevation, projecting from the intersection of $f''e''$ by the edge b''.

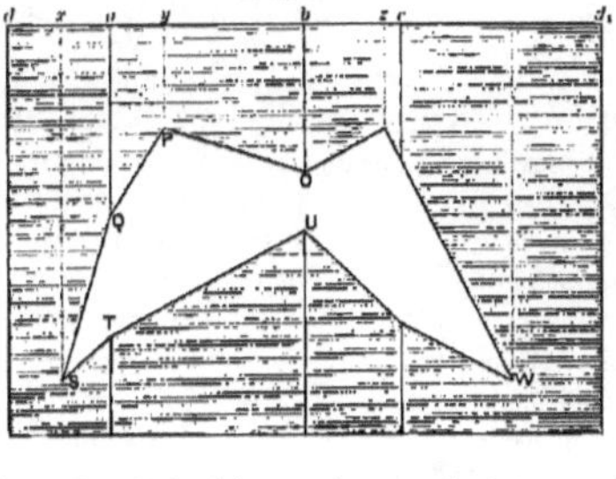

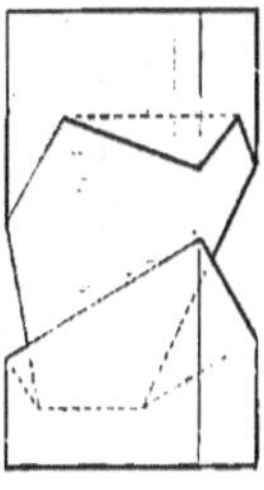

Moving from b toward w, projecting to the front elevation from either the plan or the side elevation according as we are dealing with a horizontal edge or a vertical one, we complete the intersection.

Fig. 279.

The development of the vertical prism is shown in Fig. 278. As already fully described, $d\, d_1$ = perimeter $a\, b\, c\, d$; $a\, q = a'Q'$; $b\, O = b'o'$; $a\, s = a\, s$ (of Fig. 277); $s\, S$ = vertical distance of s' from $a'e'$, etc.

160   THEORETICAL AND PRACTICAL GRAPHICS.

Although not required in shop work the draughtsman will find it an interesting and valuable exercise to draw and shade either solid after the removal of the other; also to draw the common [solid. The former is illustrated by Fig. 279; the latter by Fig. 280.

427. *The intersection of two prisms, one vertical, the other oblique but with edges parallel to V.*

Let $a\,b\,c\,d\ldots a'c'$ (Fig. 281) be the plan and elevation of the vertical prism.

Let the oblique prism be inclined 30° to H; its *faces* inclined 60° and 30° respectively to V; its base a rectangle $1\frac{1}{2}'' \times \frac{3}{4}''$, and its rear edge $\frac{3}{16}''$ back of the axis of the vertical prism.

Through some point o' of the edge $e'o'$ draw an indefinite line, $o'f'$, at 30° to H, for the elevation of the rear edge, and ff, also indefinite in length at first but $\frac{3}{16}''$ back of s, for the plan.

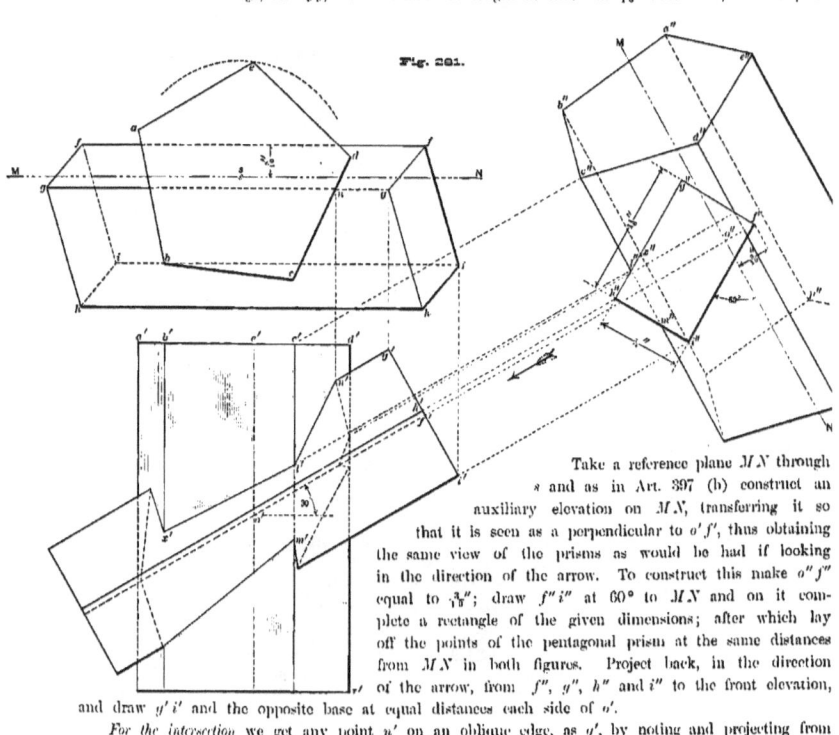

Fig. 281.

Take a reference plane MN through s and as in Art. 397 (b) construct an auxiliary elevation on MN, transferring it so that it is seen as a perpendicular to $o'f'$, thus obtaining the same view of the prisms as would be had if looking in the direction of the arrow. To construct this make $o''f''$ equal to $\frac{3}{16}''$; draw $f''i''$ at 60° to MN and on it complete a rectangle of the given dimensions; after which lay off the points of the pentagonal prism at the same distances from MN in both figures. Project back, in the direction of the arrow, from f'', g'', h'' and i'' to the front elevation, and draw $g'i'$ and the opposite base at equal distances each side of o'.

For the *intersection* we get any point n' on an oblique edge, as g', by noting and projecting from

THE INTERSECTION OF PLANE-SIDED SURFACES.

a where the plan gg meets the face cd. For a vertical edge as e'm' look to the auxiliary elevation of the same edge, as e'', getting l'' and m'' which then project back to l' and m'.

The development need not again be described in detail but is left for the student to construct, with the reminder that for the actual distance of any corner of the intersection from an edge of either prism he must look to that projection which shows the base of that prism in its true size: thus the distance of l' from the edge h' is h''l''.

428. *The intersection of pyramidal surfaces by lines and planes.* The principle on which the intersection of pyramidal surfaces by plane-sided or single curved surfaces would be obtained is illustrated by Figs. 282 and 283.

(a) In Fig. 282 the line ab, $a'b'$, is supposed to intersect the given pyramid. To ascertain s' and t'—its entrance and exit points—we regard the elevation $a'b'$ as representing a *plane* perpendicular to V and cutting the edges of the pyramid. Project m', where one edge is cut, to m, on the plan of the same edge. Obtaining n and o similarly we have mno as the plan of the section made by plane $a'b'$. The plan ab meets mno at s and t, which then project back to $a'b'$ at s' and t', the points sought.

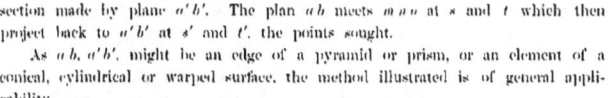
Fig. 282.

As ab, $a'b'$, might be an edge of a pyramid or prism, or an element of a conical, cylindrical or warped surface, the method illustrated is of general applicability.

(b) In Fig. 283 the auxiliary planes are taken *vertical*, instead of perpendicular to V as in the last case.

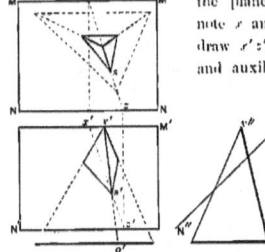
Fig. 283.

The plane MN cuts a pyramid. To find where any edge $r'o'$ pierces the plane MN pass an auxiliary vertical plane xz through the edge, and note x and z, where it cuts the limits of MN; project these to x' and z'; draw $x'z'$, which is the elevation of the line of intersection of the original and auxiliary planes, and note s', where it crosses $r'o'$. Project s' back to s on the plan of $r'o'$.

If a side elevation has been drawn, in which the plane in question is seen as a *line* $M''N''$, the height of the points of intersection can be obtained therefrom directly.

429. *The intersection of two quadrangular pyramids.* Let one pyramid be vertical; altitude $r'z'$; base $efgh$, having its longer edges inclined 30° to V.

The oblique pyramid. Let $x'y'$, the axis of the oblique pyramid, be parallel to V but inclined 6° to H, and be at some small distance (approximately ek) in front of the axis of the vertical pyramid; then xe will contain the plan of the axis, and also of the diagonally opposite edges xa and xc, if we make—as we may—the additional requirement that $a'c'$, the diagonal of the base, shall lie in the same vertical plane with the axis.

Instead of taking a separate end view of the oblique pyramid we may rotate its base on the diagonal $a'c'$ so that its foremost corner appears at b'' and the rear corner at d'', whence b' and d' are derived by perpendiculars $b''b'$ and $d''d'$, and then the edges $x'b'$ and $x'd'$. For the plans b and d use xe as the trace of the usual reference plane, and offsets equal to $b''b''$ and $d''d''$, as previously.

The angle $o'e'd'$, or ϕ, is the inclination of the shorter edges of the base to V.

The intersection. Without going into a detailed construction for each point of the outline of interpenetration it may be stated that each method of the preceding article is illustrated in this problem, and that there is no special reason why either should have a preference in any case except where by properly choosing between them we may avoid the intersection of two lines at a very acute angle—a kind of intersection which is always undesirable.

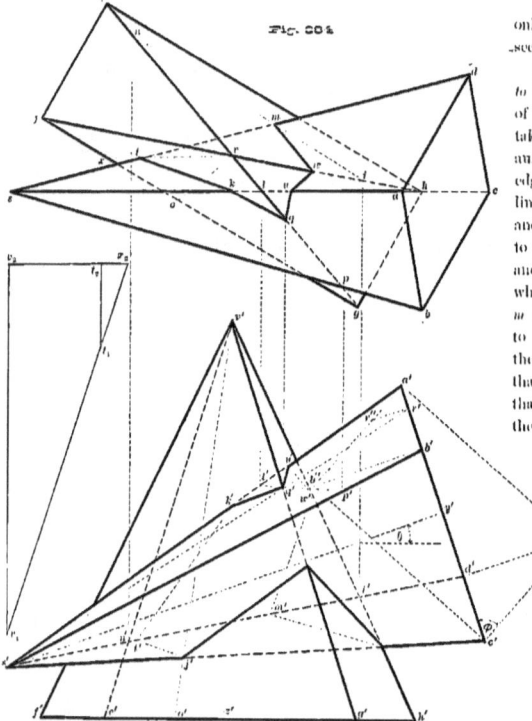

Fig. 202

In the interest of clearness only the *visible* lines of the intersection are indicated on the plan.

(a) *Auxiliary plane perpendicular to* V. To find m, the intersection of edge sd with the face ehc, take $s'd'$ as the trace of the auxiliary plane containing the edge in question; this cuts the limiting edges of the face at i' and n' which then project back to the plans of the edges at i and n. Drawing ni we note m, when it crosses sd, and project m to m' on $s'd'$. Had ni failed to meet sd within the limit of the face ehc we would conclude that our assumption that sd met that face was incorrect, and would then proceed to test it as to some other face, unless it was evident on inspection that the edge cleared the other solid entirely, as is the case with sb, $s'b'$, in the present instance. By using $s'b'$ as an auxiliary plane the student will get a graphic proof of failure to intersect.

(b) *Auxiliary plane vertical.* This case is illustrated by using rg as the trace of an auxiliary vertical plane containing the edge rg, $r'g'$. Thinking this edge may possibly meet the face sba we proceed to test it on that assumption.

The plane rg crosses sa at l, and sb at p; these project to l' on $s'a'$ and to p' on $s'b'$; then $p'l'$ meets $r'g'$ at q', which is a real instead of an imaginary intersection since it lies between the actual limits of the face considered. From q' a vertical to rg gives q.

The order of obtaining the points. The start may be with *any* edge, but once under way the progress should be uniform, and each point joined with the preceding as soon as obtained. Thus, supposing that q' was the point first found, a look at the plan would show that the edge sa of

THE INTERSECTION OF SINGLE CURVED SURFACES.   163

the oblique pyramid would be reached before eh on the other, and the next auxiliary plane would therefore be passed through sa to find uu'; then would come eh and sd. Running down from m on the face sde we find the positions such that inspection will not avail, and the only thing to do is to try, at random, either a plane through eh or one through se; and so on for the remaining points.

The developments. No figure is furnished for these, as nearly all that the student requires for obtaining them has been set forth in Art. 396, Case 6. The only additional points to which attention need be called are the cases where the intersection falls on a *face* instead of an *edge*. For example, in developing the *vertical* pyramid we would find the development of j' by drawing $e'j'$, prolonging it to o', and projecting the latter to o, when fxo would be the real distance to lay off from f on the development of the base; then laying off *the real length* of $e'j'$ on $e'o'$ as seen in the development we would have the point sought. Similarly, with tt', draw ex; make $r_2 x_2 = ex$, and $r_2 r_1 =$ altitude $e'z'$; then $r_1 x_2$ is the true length of ex (in space); also, making $r_2 t_2 = et$ and drawing $t_2 t_1$, we find $t_1 t_1$ to lay off in its proper place on the development of the same face efg.

430. *An elbow or T-joint, the intersection of two equal cylinders whose axes meet.* Taking up curved surfaces the simplest case of intersection that can occur is the one under consideration, and which is illustrated by Fig. 285.

The conditions are those stated in Art. 428 for a *plane* intersection, which is seen in $a'b'$ and is actually an ellipse.

The vertical piece appears in plan as the circle mq. To lay off the *equidistant* elements on each cylinder it is only necessary to divide the half plan of one into equal arcs and project the points of division to the elevation in order to get the full elements, and where the latter meet $a'b'$ to draw the dotted elements on the other.

The development of the horizontal cylinder is shown in the line-tinted figure. The curved boundary, which represents the developed ellipse, is in reality a sinusoid (Refer to Art. 171).

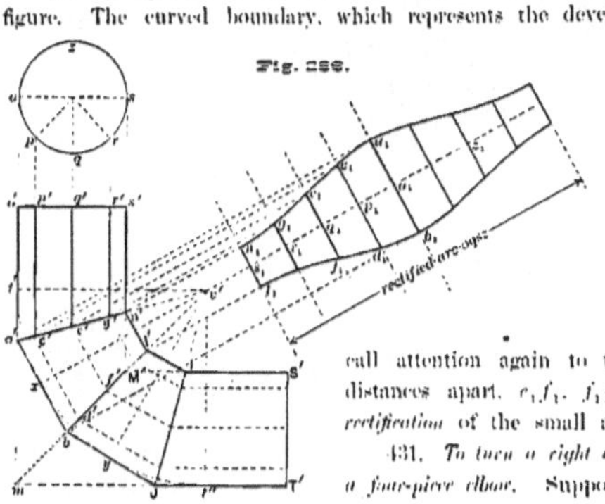

The relation of the developed elements to their originals, fully described in Art. 120, is so evident as to require no further remark, except to call attention again to the fact that their distances apart, $e_1 f_1$, $f_1 g_1$, etc., equal the *rectification* of the small arcs of the plan.

431. *To turn a right angle with a pipe by a four-piece elbow.* Supposing that the blast pipe of a furnace was to be carried around a bend by a four-piece elbow; the procedure would still closely resemble that of the last problem. Instead of *one* joint or curve of intersection there would be *three*, one less than the number of pieces in the pipe.

Let oqs show the size of the cylinders employed, and be at the same time the plan of the

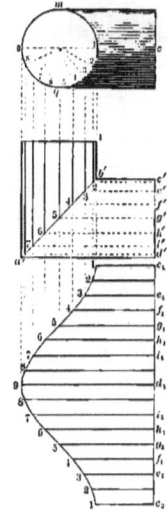

Fig. 285.

Fig. 286.

vertical piece $o'x'a'a'$. Until we know where $a'a'$ will lie we have to draw $o'a'$ and $s'n'$ until they meet the elements from S' and T', and get the joint $m M'$ as for a two-piece elbow. On $m M'$ produced take some point r', use it as a centre for an arc $t'xy t''$ tangent to the extreme elements; divide this arc, between the tangent points, into as many equal parts as there are to be pieces in the turn; then tangents at x and y—the intermediate points of division—will determine the outer limits of the joints at a', b' and J. Draw $a'r'$, finding n' by its intersection with ss'; then $n'l'$ parallel to $a'b'$, and similarly for the next piece.

The *developments* of the smaller pieces would be equal, as also of the larger. One only is shown, that of $a'n'l'b'$, laid out on the developed right section on $r's$. The lettering makes the figure self-interpreting.

432. *The intersection of two cylinders when each is partially exterior to the other.* The given condition makes it evident, by Art. 423, that a continuous non-plane curve will result.

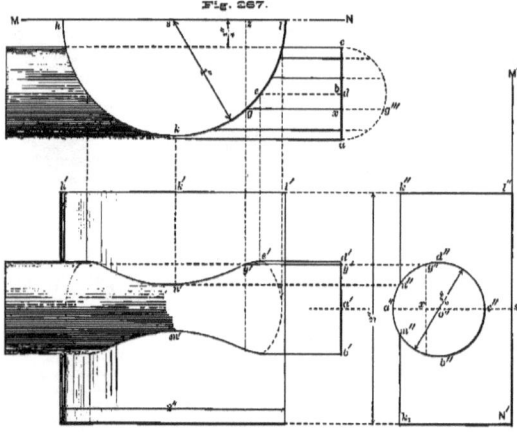

Fig. 267.

Let one cylinder be vertical, $2''$ in diameter and $2''$ high. This is shown in half plan in hkl, and in front and side elevations between horizontals $2''$ apart.

Let the second cylinder be horizontal; located midway between the upper and lower levels of the other cylinder; its diameter $\frac{4}{3}''$. On the side elevation draw a circle $a''b'' c''d''$ of $\frac{4}{3}''$ diameter, locating its centre midway between $k''l''$ and $k_1 N'$, and in such position that a'' shall be exterior to $k''k_1$. The elevation of the horizontal cylinder is then projected from its end view and is shown in part without construction lines.

The curve of intersection is obtained by selecting particular elements of either cylinder and noting where they meet the other surface.

The foremost element of the vertical cylinder is $k \ldots k'n'm'$. Its side elevation, $k''k_1$, meets the circle at a'' and m'' which give the levels of n' and m' respectively.

On the horizontal cylinder the highest and lowest elements are central on the plan and meet the vertical cylinder at c, which projects down to the elements d' and b'.

The front and rear elements, c and a_1, would be central on the elevation. The vertical line drawn from the intersection of element c with the arc hkl gives the right-hand point of the curve of intersection, at the level of a'.

Any element as gx may be taken at random, and its elevation found in either of the following ways: (a) Transfer gx, the distance of the element from MN, to $s''x$ on the side elevation, and draw xg'' and $g''g'$, to which last (prolonged) project g at g'; or (b) prolong gx to meet a

THE INTERSECTION OF SINGLE CURVED SURFACES. 165

semi-circle on ac at g'''; make $a'y' = xg'''$ and draw $y'g'$. The same ordinate xg''', if laid off *below* a, would obviously give the other element which has the same plan gx, and to which g projects to give another point of the desired curve.

433. *The intersection of a vertical cone by a horizontal prism.* Let the cone have an altitude, ww', of 4"; diameter of base, 3". (As the cylinder is entirely in front of the axis of the cone only one-half of the latter is represented.)

Fig. 236.

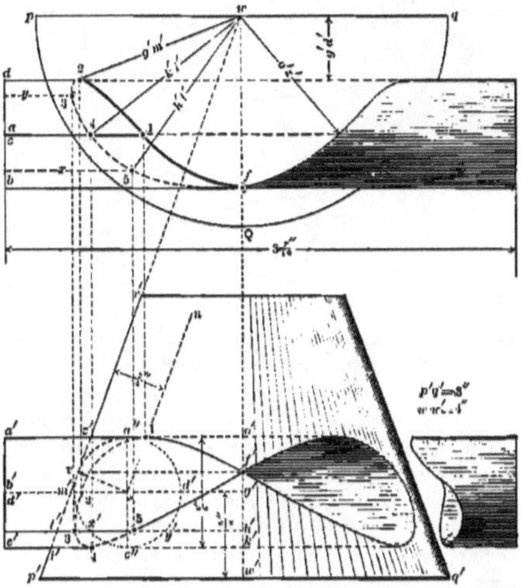

For the cylinder take a diameter of $\frac{3}{4}$"; length $3\frac{1}{16}$"; axis parallel to V, $\frac{3}{4}$" above the base of the cone, and $\frac{3}{4}$" from the foremost element. Draw ns parallel to $p'r'$ and $\frac{3}{4}$" from it; also $g'm'$ horizontal and $\frac{3}{4}$" from the base; their intersection s is the centre of the circle $a''d'c''m'$, of $\frac{3}{4}$" diameter, which bears to the element $p'r'$ the relation assigned for the cylinder to the foremost element; said circle and $p'w w'$ are thus, practically, a *side elevation* of cylinder and cone, superposed upon the ordinary view.

The dimensions chosen were purposely such as to make one element of the cone tangent to the cylinder, that the curve of intersection might cross itself and give a mathematical "double point."

The width db, of the plan of the cylinder, equals $m'd'$. The plan of the axis (as also of the highest and lowest elements, a' and c') will be at a distance sg' from w. Any element as $x'y'h'$ is shown in plan parallel to pq, and at a distance from it equal either to $h'y'$ if on the rear or to $h'x'$ if on the front.

The element through r, on which f' falls, is not drawn separately from bf in plan, since rf' and $m'g'$ are so nearly equal to each other; but f must not be considered as on the foremost element of the *cylinder*, although it is apparently so in the plan.

For the *intersection* pass auxiliary horizontal planes through both surfaces; each will cut from the cone a circle whose intersection with cylinder-elements in the same plane will give points sought. A horizontal plane through the element a' would be represented by $o'o'$, and would cut a circle of radius $o'z'$ from the cone. In plan such circle would cut the element a at point 1, and also at a point (not numbered) symmetrical to it with respect to wQ. Similarly, the horizontal plane through the element $x'h'$ cuts a circle of radius $l'h'$ from the cone; in plan such circle would meet the elements x and y in two more points (5 and 3) of the curve.

As the curve is symmetrical with respect to wQw' the construction lines are given for one-half only, leaving the other to illustrate shaded effects. The small shaded portion of the elevation of the cylinder is not limited by the curve along which it would meet the cone, but by a random curve which just clears it of the right-hand element of the cone.

434. *To find the diameter and inclination of a cylindrical pipe that will make an elbow with a conical*

166   THEORETICAL AND PRACTICAL GRAPHICS.

pipe on a given plane section of the latter. Let *cab* be a vertical cone, and *cd* the elliptical plane section on which the cylindrical piece is to fit. The diameter of the desired cylinder will equal the shorter diameter of the ellipse *cd*. To find this bisect *cd* at *e*; draw *fh* horizontally through *e*, and on it as a diameter draw the semi-circumference *fgh*; the ordinate *eg* is the half width of the cone, measured on a perpendicular to the paper at *e*, and is therefore the radius of the desired cylinder.

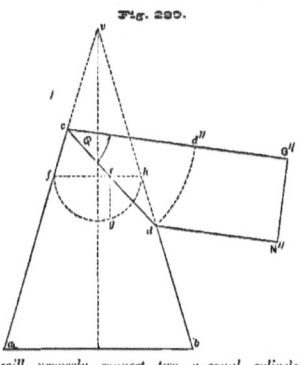

Fig. 289.

In Fig. 290, the base *NG* equals twice *ye* of Fig. 289. At first indefinite perpendiculars are erected at *N* and *G*, on one of which a point *C* is taken as a centre for an arc of radius equal to *cd* in Fig 289. The angle φ being thus determined is next laid off in Fig. 289 at *c*, and *cdN″G″* made the exact duplicate of *CDNG*, completing the solution.

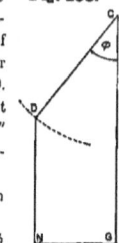

Fig. 290.

The developments are obtained as in Arts. 120 and 191.

435. *To determine the conical piece which will properly connect two unequal cylinders of circular section, whose axes are parallel, meeting them either* (a) *in circles or* (b) *in ellipses; the planes of the joints being parallel.*

(a) *When the joints are circles.* To determine the conical frustum *bchc* prolong the elements *eb* and *hc* to *r*; develop the cone *r...eh* as in Art. 418, and on each element as seen in the development lay off the real distance from *r* to the upper base *bc*. Thus the element whose plan is *c₁k* is of actual length *rk₁* and cuts the upper base at a distance *ru* from the vertex, which distance is therefore laid on *rk₁* wherever the latter appears on the development.

(b) *When the joints are ellipses.* Let the elliptical joints *ao* and *qr* be the bases of the conical piece *quor*. To get the development complete the cone by prolonging *qn* and *or* to *w*; prolong *qr* and drop a perpendicular to it from *w*; find the minor axis of the ellipse *qr* as in the first part of Art. 434 and having constructed the ellipse proceed as in Art. 418, since in Fig. 255 the arc *abc...g* is merely a special case of an ellipse.

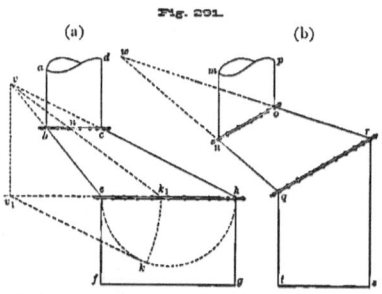

Fig. 291.

436. *The projections and patterns of a bath-tub.* Before taking up more difficult problems in the intersection of curved surfaces one of the most ordinary applications of Graphics is introduced, partly by way of illustrating the fact that the engineer and architect enjoy no monopoly of practical projections.

In Fig. 292 the height of the main portion of the tub is shown at *a′d′*. Let it be required that the head end of the tub be a portion of a *vertical right cone* whose base angle *c′b′a′* equals the flare of the sides, such cone to terminate on a curve whose vertical projection is *o′n′z′a′*. Draw

THE INTERSECTION OF SINGLE CURVED SURFACES. 167

two lines, $b'l'$ and $c'i'$, at first indefinite in length and at a distance $a'd'$ apart. Take $a'd'$ vertical, and regard it not only as the projection of the elements of tangency of the flat sides with the conical end, but also as the elevation of part of the axis, prolonging it to represent the latter. Use e, the plan of the axis, as the centre for a semicircle of radius rc, whose diameter ed is the width of the bottom of the tub. Project e to e'; make angle $e'c'd'$ equal to the predetermined flare of the sides; prolong $e'c'$ to b' and o'; project b' to b on ec prolonged and draw are abm with radius br, obtaining am for the width of the plan of the top.

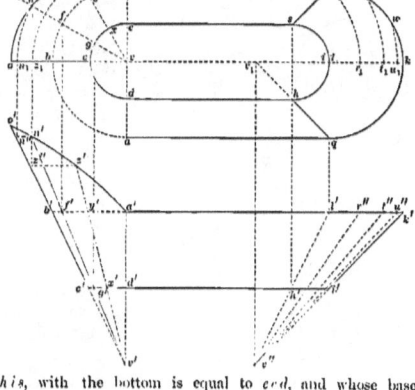

Fig. 292.

The plan of one-half the curve $o'n'z'a'$ is shown at $onzm$ and is thus found: Assume any element $e'x'y'$; prolong it to z'; obtain the plan exy and project z' upon it at z. Similarly for n and as many intermediate points as it might seem desirable to obtain.

Assuming that the foot of the tub is composed of an *oblique* cone whose section, his, with the bottom is equal to ccd, and whose base angle is $h'i'k'$, we project i to i', draw $i'k'$ at the given angle to the base, project k' to k, and through the latter draw the semicircle rkq with radius br, obtaining the plan of the upper base.

Joining the tangent points r and s, h and q, we have rs and hq as the elements of tangency of sides with end. Their elevations coincide in $h'l'$, which meets $k'i'$ at r'', whose plan is r_1 on hq.

Fig. 293.

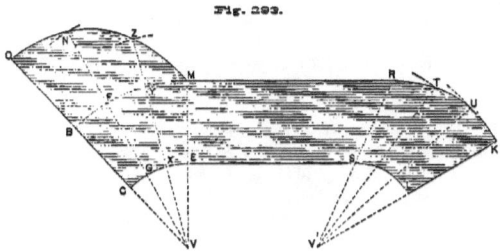

The development. Fig. 293 is the development of one-half of the tub. EM equals $b'c'$; VO equals $v'o'$; VZ equals $v'z''$, the true length of $v'z'$, obtained, as in previous constructions, by carrying z to z_1, thence to level of z'. Similarly at the other end. (Reference Articles 191, 408, 418.)

487. *The intersection of a vertical cylinder and an oblique cone, their axes intersecting.*

Let MBd and $M'R'P'N'$ be the projections of the cylinder; $c', a'b'$ and c, a, n, b, m those of the cone. The axes meet o' at an angle θ which is arbitrary.

The ellipse *a n b m* is supposed to be constructed by one of the various methods employed when the axes are known; and in this case we get the *length* of *m n* from *a' b'* and its *position* from *n'*, while *a b* is vertically above *a' b'*.

(a) *Solution by auxiliary vertical planes.* Any vertical plane *r l s* will cut elements from the cylinder at *e* and *l*; also, from the cone, elements which meet the base at *s* and *t*. Project *s* and *t* to *s'* and *t'*, join the latter with the vertex *v'* and note *l'* and *e'* (just below *d'*) where they cross the vertical projection of the elements from *l* and *e*; these will be points in the desired curve of intersection.

By assuming a sufficient number of vertical planes through *r* the entire curve can be determined.

(b) *Solution by auxiliary spheres.* If two surfaces *of revolution* have a common axis they will intersect each other in a circle whose plane is perpendicular to that axis.* This property can be advantageously applied in problems of intersection.

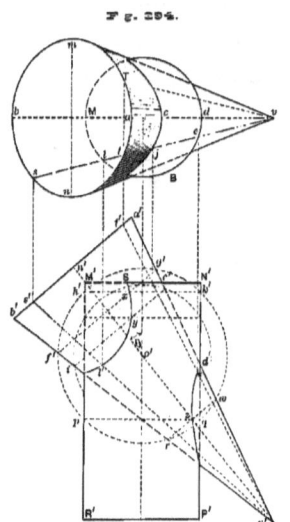

Fig. 294.

With *o'*—the intersection of the axes—as a centre, we may draw circles with random radii *o'f'*, *o'i*, and let these represent spheres. The sphere *f'g'w* intersects the *cone* in the circle *f'g'*; the *cylinder* in the circle *h'k'*. These circles intersect each other at *x* in a common chord whose extremities are points of the curves sought. They are both projected in the point *x*.

A second pair of circular sections, lying on the same auxiliary sphere, are seen at *p q* and *r w*, their intersection *z* being another point in the solution. The point *y* results from taking the smaller sphere.

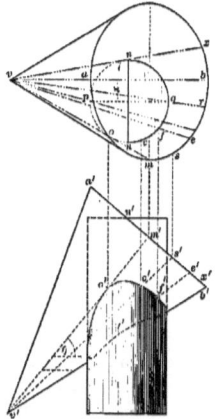

Fig. 295.

438. *Intersection of a cylinder and cone, their axes not lying in the same plane.*

In Fig. 295 let the cylinder be vertical and the cone oblique, the axis of the latter being parallel to V and inclined θ° to H, and also lying at a distance *x* back of the axis of the cylinder.

The auxiliary surfaces employed may preferably be vertical planes through the vertex of the cone, since each will then cut elements from both cylinder and cone. Thus, *r f c* is the h. t. of a vertical plane which cuts *e r, e'r'* from the cone, and the vertical element through *f* from the cylinder; these meet in vertical projection at *f'*, one point of the desired curve. The *plan* of the intersection obviously coincides with that of the cylinder.

---

*By the definition of a surface of revolution (Art. 340) any point on it can generate a *circle* about its axis. If, then, two surfaces have *the same axis*, any point common to both surfaces would generate one and the same circle, which must also lie on both surfaces and therefore be their line of intersection.

THE INTERSECTION OF SINGLE CURVED SURFACES. 169

439. *Conical elbow; right cones meeting at a given angle and having an elliptical joint.* This is one of the cases mentioned in Art. 423 as not admitting of illustration in the same way as when dealing with surfaces of uniform cross section, but a plane intersection is nevertheless secured as with cylinders by making the *extreme elements* of the cones intersect.

Let vx in Fig. 296 be the axis of one of the cones. If xyz is the required angle between the axes bisect it by the line ym, and draw the joint cd parallel to such bisector. The right cone which is to meet $abcd$ on cd must be capable of being cut in a section equal to cd by a plane making an angle θ with its axis, and must obviously have the same base angle as the original cone; since, however, the upper portion vdc of the given cone fulfills

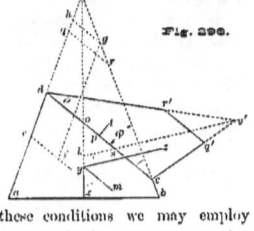

Fig. 296.

these conditions we may employ it instead of a new cone, rotating it about an axis pt which is perpendicular to the plane of the ellipse dc and passes through its centre. The point o, in which the axis vx meets the plane dc, will then appear at s, by making $op = ps$; sv', drawn parallel to yz, will be the new direction of vo; and an arc from centre d with radius cv will give v', which is then joined with d and c to complete the construction.

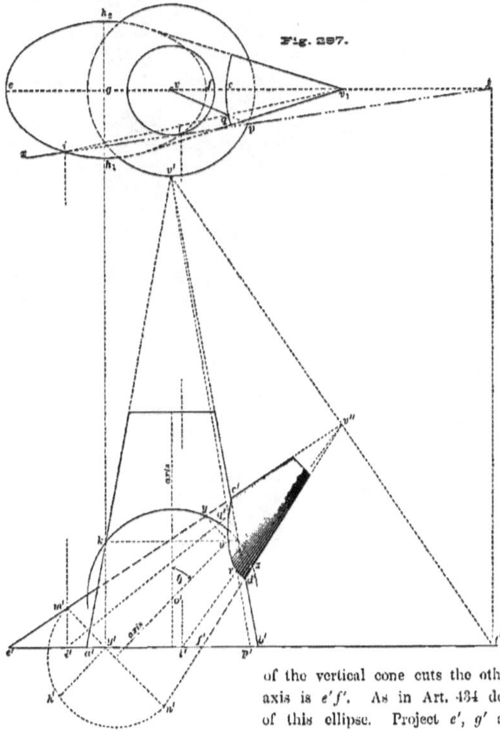

If the length of the major axis of the elliptical joint had been assigned, as ef for example, that length would have first been laid off from some point e on the extreme element and parallel to ym, then from f a parallel to vc, giving g on vc; then gh, parallel and equal to ef, gives the joint in its proper place.

440. *Right cones intersecting in a non-plane curve; axes meeting at an oblique angle.* Let one cone, $v'.a'b'$, (Fig. 297) be vertical; the other, oblique, its axis meeting $v'o'$ at an angle θ.

Fig. 297.

The plane $a'b'$ of the base of the vertical cone cuts the other cone in an ellipse whose longer axis is $e'f'$. As in Art. 434 determine $g'h'$, the semi-minor axis of this ellipse. Project e', g' and f' up to e, g and f; make

170   THEORETICAL AND PRACTICAL GRAPHICS.

gh_1 and gh_2, each equal to $g'h'$; then on ef and h_1h_2 as axes construct the ellipse eh_1fh_2 as in Art. 131. Tangents from r_1 to the ellipse complete the plan of the oblique cone.

(a) *The curve of intersection, found by auxiliary planes.* In order that each auxiliary plane shall contain an element (or elements) of each cone, it must contain both vertices and therefore the line $v'v''$, which joins them; hence its trace on the plane $e'a'b'$ must pass through the trace, $t't$, of such line on that plane. Take tx as the horizontal trace of one of these auxiliary planes. It cuts elements starting at i and l on the base of the oblique cone. One of the elements cut from the other cone is rp, which in vertical projection ($e'p'$) crosses the elevations of the other elements at q' and r', two points of the curves sought. Since the extreme elements of the cones are parallel to V we will have c' and d'—the intersections of their elevations—for two more points of the curve. Having found other points by repeating the same process the curve $c'q'rd'$ is drawn through them, and the cones may then be developed as in Art. 191.

Fig. 288.

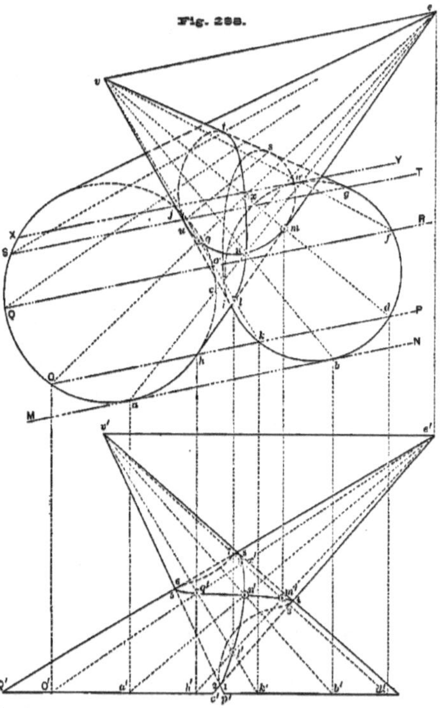

(b) *Method by auxiliary spheres.* Since the axes intersect we may use auxiliary spheres as in Case (b) of Art. 437. Thus, with o'—the intersection of the axes—as a centre, take any radius $o'k$ and regard arc kyz as representing a portion of a sphere which cuts the cones in ks and yz. These meet at w, one point of the curve of intersection $c'q'd'$.

441. *Intersecting cones, bases in the same plane but axes not.* Let $r.kbfy$ and $e.sQhj$ be the plans of the cones; $v.'p'd'$ and $e.'Q'e'$ their elevations.

As argued in Case (a) of the last problem, the auxiliary planes must contain the line joining the vertices; their H-traces would therefore, in the general case, pass through the trace of that line upon the plane of the bases; but, in the figure, both vertices having been taken at the same height above the bases, the line which joins them must be *horizontal*, hence *parallel* to the H-traces of the auxiliaries: that is, XY, ST, QR, etc., are *parallel* to vc.

It happens that the trace MN of the foremost auxiliary plane is *tangent* to both bases, hence contains but one element of each cone and determines but one point of the desired curve. These elements, ac and bc, meet at n, while their elevations intersect at n'.

THE INTERSECTION OF SINGLE CURVED SURFACES. 171

Each of the other planes, except XY, being secant to both bases, will cut two elements from each cone, their mutual intersections giving four points of the curve of interpenetration. Thus, in plane OP, the element Oe meets ck in q and vd in x, while element he gives l and m on the same elements.

The plane XY being tangent to one base while secant to the other gives but two points on the curve sought.

Order of connecting the points. Starting with any plane, as MN, we may trace around the bases either to the right or left. Choosing the former we find, in the next plane, the point h to the right of a on one base, and d similarly situated with respect to b on the other; therefore m, on he and dv, is the next point to connect with a. Elements ue and fv give the next point, then ue and gv locate s, after which those from j and w give the last before a return movement on the base of the v-cone. As nothing new would result from retracing the arc gfd we continue to the left from w, although compelled to retrace on the other base, since planes beyond j would not cut the v-cone. The element ue is therefore taken again, and its intersection noted with an element whose projection happens to be so nearly coincident with vx that the latter is used.

Continuing along arcs oeh and ikb we reach the plane MN again, the curves ilx and qnm crossing each other then at n—the point lying in that plane. Such point is called a *double point*, and occurs on non-plane curves of intersection at whatever point of two intersecting surfaces they are found to have a common tangent plane.

Tracing to the left from a and to the right from b the elements Oe and dv are reached, in the plane OP. Their intersection x is joined with n on one side and with the intersection of Se and gv on the other. Soon the tangent plane XY is again reached and a return movement necessitated, during which the arc $NSQOa$ is retraced, while on the other base the counter-clockwise motion is continued to the initial point b, completing the curve.

Visibility. The visible part of the intersection in either view must obviously be the intersection of those portions of the surfaces which would be visible were they separate, but similarly situated with respect to H and N.

In plan the point n lies on visible elements, and either arc passing through it is then visible till it passes (becomes tangent to, in projection) an element of extreme contour as at m or t, when it runs from the upper to the under side of the surface and is concealed from view.

The point w would be visible on the v-cone but for the fact that it is on the under side of the c-cone.

A similar method of inspection will determine the visible portions of the vertical projection of the curve, which will not be identical with those of the plan. In fact, a curve wholly visible in one view might be entirely concealed in the other.

442. *The intersection of a vertical cylinder and an oblique cone, their axes in the same plane.* If in Art. 440 the vertex v' were removed to infinity the v-cone would become a vertical cylinder; the line $v'v''$ would become a vertical line through v''; t would be vertically above v''; but the *method* of solving would be unchanged.

443. In general, any method of solving a problem relating to a cone will apply with equal facility to a cylinder, since one is but a special case of the other. The line, so frequently used, that passes through the vertex of a cone in the one problem is, in the other, a parallel to the axis of the cylinder. Planes containing both vertices of cones become planes parallel to both axes of cylinders.

In view of the interchangeability of these surfaces it is unnecessary to illustrate by a separate figure all the possible variations of problems relating to them.

172   *THEORETICAL AND PRACTICAL GRAPHICS.*

444. *Intersection of two cones, two pyramids, or of a cone and a pyramid, when neither the bases nor axes lie in one plane.*

One method of solving this problem has been illustrated in Art. 429, where the intersection was found by using auxiliary planes that were either vertical or perpendicular to V; we may as easily, however, employ the method of the last problem, viz., by taking auxiliary planes so as to contain both vertices. This will be illustrated for the problems announced, by taking a cone and pyramid; and, for convenience, we will locate the surfaces so that one of them will be vertical, and the base of the other will be perpendicular to V, since the problem can always be reduced to this form.

Fig. 299.

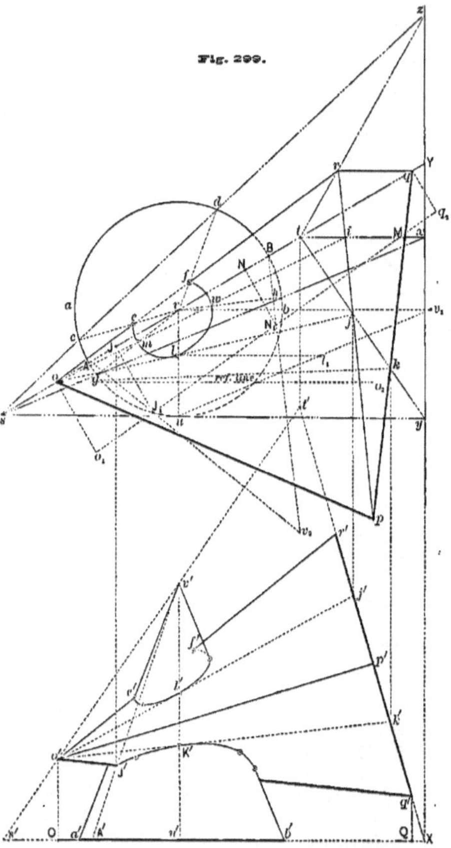

Let the cone $v'. a'b'$, $v.cdB$, (Fig. 299) be vertical, and the pyramid $o'.r'q'p'$, $o.rqp$, inclined.

We will assume that the projections of the pyramid have been found as in preceding problems, from assigned data, using oo_2, $o'p'$, (taken perpendicular to the base $r'q'$) as the reference line.

Join the vertices by the line $v'o'$, vo, and prolong it to get its traces, ss' and tt', upon the planes of the bases. All auxiliary planes containing the line vo, $v'o'$, must intersect the planes of the two bases in lines passing through such traces.

Prolong $r'q'$ to meet the plane $a'b'$ at X. Project up from X, getting yz for the plan of the intersection of the two bases.

We may assume any number of auxiliary planes, some at random, but others more definitely, as those through edges of the pyramid or tangent to the cone. Taking first one through an edge, as or, we have trz for its trace on the pyramid's base, then zs for its trace on H. The elements cv and dv which lie in this plane meet the edge or at e and f, giving two points of the curve. These project to $o'r'$ at e' and f'.

INTERSECTIONS.—PRACTICAL ENGINEERING DESIGNS. 173

The plane sy, tangent to the cone along the element av, has the trace yt on the base of the pyramid, and cuts lines jo and ko from its faces. These meet vu at two more points of the curve, their elevations being found by projecting j to j' and k to k', drawing $o'j'$ and $o'k'$, and noting their intersections with $r'u'$. To check the accuracy of this construction for either point, as l, draw rr_1 perpendicular to vu and equal to $v'u'$, join r_1 with u, and we have in vv_1u the rabatment of a half section of the cone, taken through the element vu and the axis; then ll_1, parallel to rr_1, will be the height of l' above the base $a'b'$.

With one exception, any auxiliary plane between sy and sz will give four points of the intersection. The exception is the plane sY, containing the edge oq, and which, on account of happening to be vertical, requires the following special construction if the solution is made wholly on the plan: Rabat the plane into H; the elements it contains will then appear at Av, and Bv_1, while the edge uq will be seen in o_1q_1 (by making $ou_1 = o'O$, and $qq_1 = q'Q$); elements and edge then meet at J_1 and N_1 which counter-revolve to J and N. We might, however, get elevations first, as J', by the intersection of element $A'v'$ with edge $o'q'$; then J from J'.

In the interest of clearness several lines are omitted, as of certain auxiliary planes, hidden portions of the ellipses, and the curves in which srq (the rear face) cuts the cone. The student should supply these when drawing to a larger scale.

BRIDGE POST CONNECTIONS.—GEARING.—SPRINGS.—BOLTS, SCREWS AND NUTS.

445. *Detail of a Bridge.*—*Upper-Chord Post-Connection.*—A bridge or roof truss is an assemblage of pieces of iron or wood, so connected that the entire combination acts like a single beam. Figs. 300 and 301 are what are called "skeleton diagrams" of bridge trusses, each piece or "member" of the truss being represented by a single line. $ABCD$ and $A'B'C'D'$ are the trusses proper, the

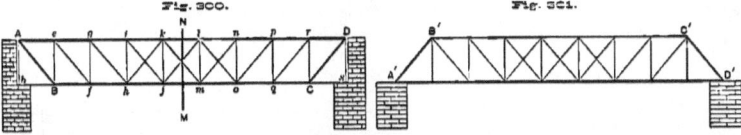

Fig. 300.   Fig. 301.

former being for an overhead track and the latter for a roadway running through the bridge. In each case the upper part—called the *upper chord*—(AD, $B'C'$) sustains *compression*, and is made of "built beams," formed by riveting together various plates and lengths of structural iron in such manner as to form one practically continuous column.

The *lower chords* (BC, $A'D'$) sustain tension, and are made of bars of high tensile strength.

The members that connect the chords are called either *ties* or *struts* according as the strain in them is tensile or compressive. Collectively they form the *web* of the truss.

In the form of truss illustrated—which is only one of many which have commended themselves to the profession—the vertical pieces or "posts," Bv, fg, etc., sustain compression, and are therefore "built" columns. They divide the trapezoid into parts called *panels*, which has given the name *panel system* to this largely-employed arrangement of bridge members.

All the diagonal members in both figures, excepting $A'B'$ and $C'D'$, are tension bars or rods. Bb and Cs are struts whose sole office is to keep the posts Ab and Ds vertical; said posts then conveying to the masonry whatever weights are transmitted through the truss to A and D respectively.

174   THEORETICAL AND PRACTICAL GRAPHICS.

Fig. 302 is a perspective view of the connection at A between the post Ab, the upper chord AD, and the four diagonal bars that are projected in AB. The working drawings required for such connection are shown in the three views on the opposite page. Three analogous views would be required for the connection at the foot of the same post.

Fig. 302.

When—as in the case from which our example is taken—there are *two* railroad tracks overhead, the members of the middle truss will usually have different proportions from those in the outside trusses, and a separate set of three views has therefore to be made for each of its post connections, so that the smallest number of shop drawings for one such bridge—after making all allowance for the symmetry of the structure with reference to the central plane MN—would consist of twenty such groups of three as are illustrated by Fig. 304.

The upper projections (Fig. 304) are obviously a front and a side elevation. The lower figure may preferably be regarded as a plan of the object inverted, since that conception is somewhat more natural than that of the post in its normal position, while the draughtsman lies on his back and gazes up at it from beneath.

Fig. 303 shows the inverted plan on a somewhat smaller scale, and, although presented mainly to illustrate the contrast between views with shade lines and without, contains one or two serviceable dimensions that are omitted on the other plate.

446. *General description.* Referring to the wood cut as well as the orthographic projections, we find the upper chord to be composed of a long cover-plate, $18'' \times \frac{7}{16}''$, riveted to the top angles of two vertical channel bars set back to back; each channel being 15" high and weighing 200 pounds per yard. The cross section of the upper chord is shown in solid black, with just enough space intended between plate and channels to show that they are not all in one piece.

Perpendicular to the vertical faces of the channels and through holes cut therein runs a cylinder called a "pin," 4" in diameter and 21$\frac{3}{4}$" "between shoulders" (as marked on the plan), that is, between the planes where the diameter is reduced and a thread turned, on which connection can be made with the corresponding post in the next truss.

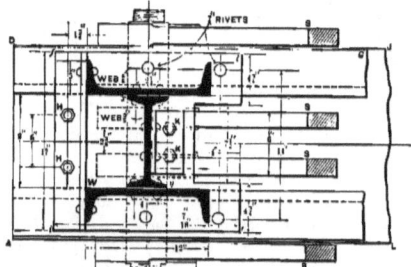

Fig. 303.

Four diagonal bars are sustained by the pin, the latter passing through holes, called "eyes," in the bars. Two of the bar heads are between the channels.

Two plates are inserted between each of the outside bars and the nearest channel, not only to prevent the bar from touching the angle, as at h, but also to relieve the metal nearest the pin from some of the strain. The longer plate $m n F E$ is next the channel. The other, $n m o p$, has a kind of hub cast on it which rounds up to the bar head, as shown in the side elevation.

The vertical post is made up of an I-beam and two channels, as shown by the black sections on the plan.

UPPER-CHORD POST-CONNECTION. 175

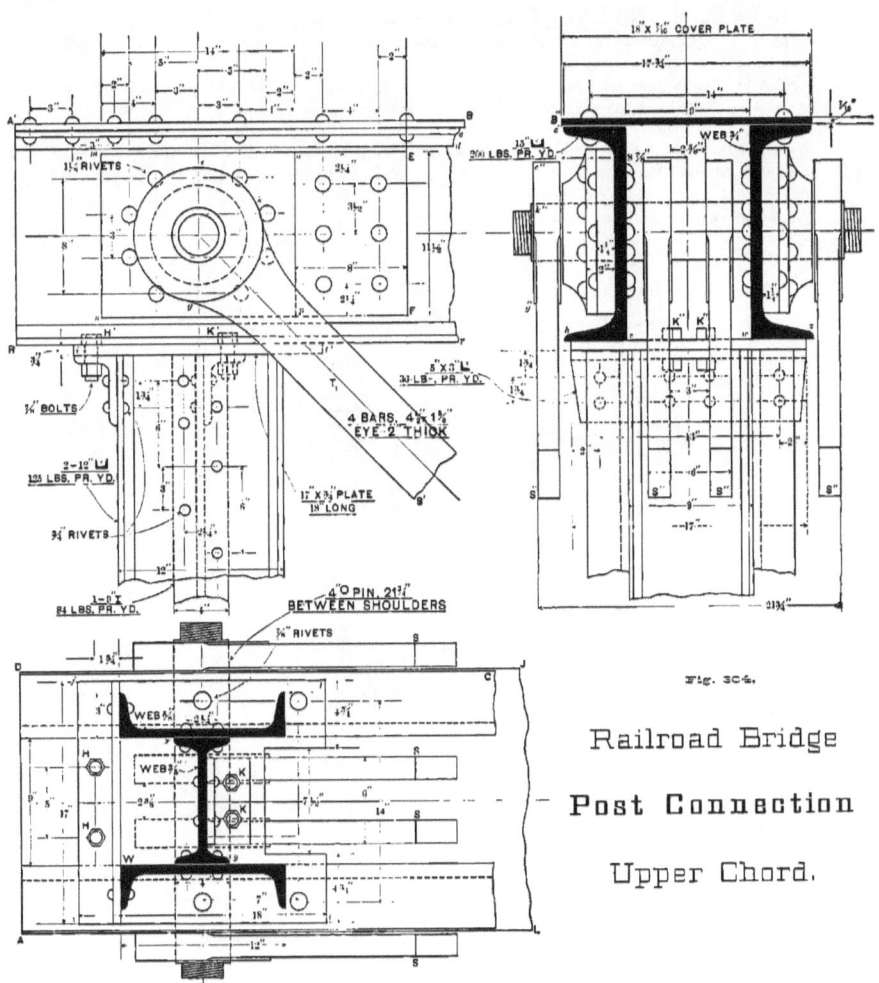

Fig. 304.

Railroad Bridge
Post Connection
Upper Chord.

When drawing the above the student should make *horizontal* dividing lines in all fractions. A brush tint of Prussian blue for all the metal parts will enhance the appearance of the drawing very materially, but the previous lining should be, obviously, in best waterproof ink. Centre, dimension, and extension lines should be in continuous red lines, unless for blue printing, in which case all lines will be black.

176   THEORETICAL AND PRACTICAL GRAPHICS.

Between the upper chord and the top of the post is a three-quarter-inch plate, seen best on the plans at *i f l t*. It is nicked out 4", near the nuts *K*, so as to clear the two middle bars *S* which come between the channels.

A 5" × 3" angle-iron runs from outside to outside of channels, and is held by rivets and by the bolts marked *H*. A shorter piece of the same kind is fastened by bolts *K* to the plate, and by rivets to the web of the I-beam.

447. *Hints as to drawing the bridge post connection.* Draw the main centre lines first; then the plan and side elevations simultaneously, as the horizontal centre line of the plan represents the same vertical reference-plane as the vertical centre line of the side elevation, and one spacing of the dividers may be made to do double work.

The solid sections should be drawn first of all; then the pins, bars, and cap plates of the post in the order named. The parts already drawn should next be represented on the front elevation by

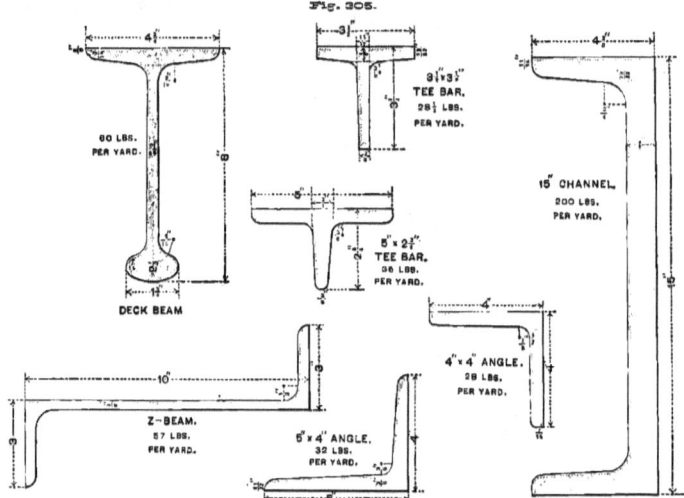

Fig. 305.

projecting up from the plan and across from the side view. The filler plates, *m p* and *m F*, with their rivets, come next on the front elevation, from which they project to the side.

Next in order draw the angle irons on the front elevation, with their bolts, *H'* and *K'*, and project them both across and down. Finally put in all remaining rivets, and dimension the views.

The angle whose bolts are marked *H* terminates exactly on the edges of the channels, as shown in the wood-cut, rather than as indicated in the side elevation.

448. *Structural Iron.* In Figs. 305 and 306 the forms of iron more generally used in bridge and house construction are shown in cross section, and may advantageously be drawn on an enlarged scale.

Treated as described in Art. 75 they may be worked up with brush or pen like Fig. 136.

TOOTHED GEARING.—HELICAL SPRINGS.

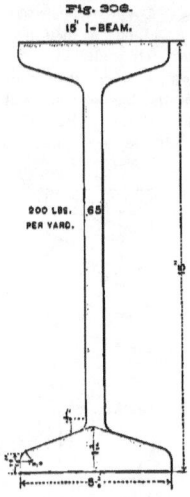

Fig. 306.
15" I-BEAM.
200 LBS. PER YARD.

449. *Toothed Gearing.* When two shafts are to be rotated and a constant velocity ratio maintained between them, it is customary to fix upon them toothed wheels whose teeth are so proportioned that by their sliding action upon each other they produce the motion desired. It is not within the intended scope of this work to go at length into the theory of gearing, for which the student is referred to such specialized treatises as those of Grant, Robinson, MacCord, Weisbach and Willis; but the draughtsman will find it to his advantage to be familiar with the following rapid method of drawing the outlines of the teeth of a spur wheel, in which a remarkably close approximation is made by circular arcs to the theoretical involute outlines now so much employed.

450. CN (in Fig. 307) is the radius of the *pitch circle*, that is, the circle which passes through the middle of the working part of the tooth.

The working outlines outside the pitch circle are called *faces* (fg, hi), while *within* they are designated as *flanks*. The flanks are rounded off into the *root circle* by small arcs called *fillets*.

The limits of the teeth on the addendum circle, as a, g, h, m, are called their *points*.

On the pitch circle the distance bi, between corresponding points of consecutive teeth, is called the *circular pitch* (usually denoted by P).

Knowing the pitch and the number (N) of teeth, the radius of the pitch circle will equal $P \times N \div 2\pi$.

As one inch pitch and twenty teeth are taken as data for the illustration, we have $CN = 3''.18 +$.

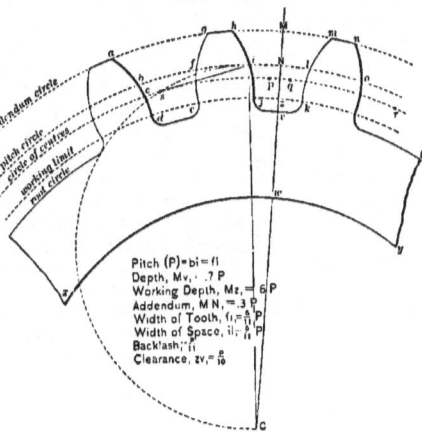

Fig. 307.

Pitch (P)=bi=fl
Depth, Mv, .7 P
Working Depth, Mz, = 6/P
Addendum, M N, = .3 P
Width of Tooth, G, = 5/11 P
Width of Space, il = 6/11 P
Back'ash;"il
Clearance, zv, = 2/10

The other proportions are also expressed in terms of the pitch, a frequently-used system therefor being indicated on the figure.

If i is a point through which a tooth outline is to pass, draw Ci, and on it as a diameter describe the semi-circumference Csi. An arc from centre i, with a radius of one-fourth Ci, will give the centre s of the outline hij.

Draw the "circle of centres" through s, from centre C. Then with si in the dividers, and from centre f find q, which use in turn for arc gfe, and so continue.

The *width of rim*, vw, is often made, by a "shop" rule, equal to three-fourths the pitch. Reuleaux gives for it the following formula: $vw = 0.4 P + .12$.

Diametral pitch is very frequently used instead of circular pitch, and is simply the number of teeth per inch of pitch-circle diameter.

451. *Helical Springs.* Draw first (Fig. 308) a central helix $acfm..T$, as follows: Divide aa_1—

178   *THEORETICAL AND PRACTICAL GRAPHICS.*

which is the *pitch*, or rise in one turn — into any number of equal parts, and the semi-circumference *AEM* into half as many equal divisions; then each point marked with a capital on the half plan gives two elevations (denoted by the same letter small) by a process which is self-evident.

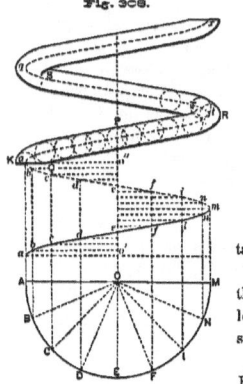

Fig. 308.

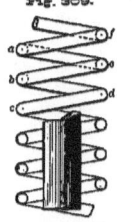

Fig. 309.

If the spring is circular in cross-section draw a series of circles having centres on the helix, and whose diameters equal that of the spring; then the outlines of the spring will be curves that are tangent to the circles.

Fig. 310.

If the spring be small the curvature of the helix may be ignored, and a series of parallel straight lines employed instead, drawn tangent to circular arcs as in Fig. 309. The upper half of the figure gives the method of construction, while the lower shows the spring in section, and surrounding a solid cylindrical core.

452. *Springs of rectangular cross-section.* Fig. 311 shows a spring of this description, formed by moving the rectangle *a b c d* helically, each point describing a helix which can be constructed as described in the last article.

When any considerable number of turns of the same helix has to be drawn it will save time if the draughtsman will shape a strip of pear-wood into a *templet*, i. e., a piece whose outline conforms to a line to be drawn or an edge to be cut, using it then as a curved ruler to guide his pen. This is the preferable method for all large work.

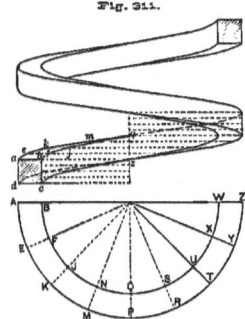

Fig. 311.

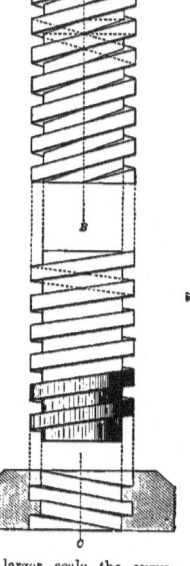

453. *Square-threaded screws.* If instead of spirally twisting a rectangular bar the same kind of surface be cut upon a cylinder of wood or metal, we shall have a square-threaded screw. This is illustrated by the upper part of Fig. 310, and its construction is self-evident after what has preceded. On a larger scale the curvature of the helices would have to be indicated.

The upper view is an elevation of a small *double-threaded* square screw, generated by winding two equal rectangles simultaneously around the axis.

The central figure is an elevation of a *single-threaded* screw. The lower figure is a sectional view of the nut for the single-threaded screw, and evidently presents a surface identical with that of the back half of the screw which fits it.

454. *Triangular-threaded screws.—United States Standard.*—The proportions devised by Mr. William Sellers of Philadelphia have been so generally adopted as to be known as the United States Standard. They are given in the table on the next page.

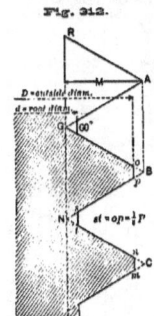

Fig. 312 shows a section of the Sellers screw. It is blunt on the thread, and also at the root. The part opB which is removed from the point may be regarded as filled in at Nst. AB being the pitch (P), the widths op, st, are each one-eighth of P.

With N equal to the number of threads per inch, and D the outside diameter of the screw or bolt, the value of d—the diameter at the root—may be obtained from the formula $d = D - (1.299 \div N)$.

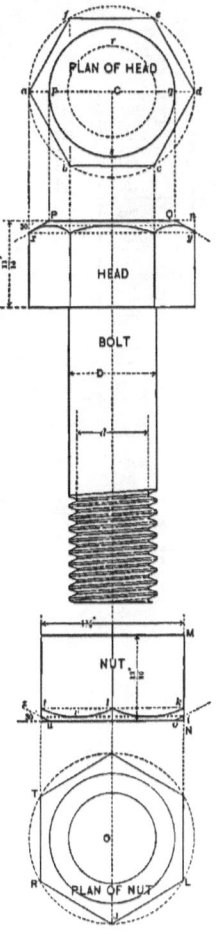

Other proportions are as follows: *The pitch* is equal to $0.24\sqrt{D+0.625} - 0.175$. *The depth of thread* equals $0.65 P$. For bolts and nuts, whether hexagonal or square, the "width across flats," or shortest distance between parallel faces, equals $1.5 D$, plus one-eighth of an inch for rough or unfinished surfaces, or plus one-sixteenth of an inch for "finished," i. e., machined or filed to smoothness.

The depth of nut equals the diameter of the bolt, for "rough" work. Tables should be consulted for the proportions of finished pieces.

Fig. 313 is a drawing, to reduced scale, of a finished $\tfrac{3}{4}''$ bolt. The elevations show a bevel or chamfer, such as is usually given to a finished bolt or nut. On the plans this is indicated by the circles of diameter pq, the latter usually a little more than three-fourths of the diameter ad.

To draw the lines resulting from chamfering proceed thus: On a view showing "width across flats," as that of the nut, draw the chamfer lines zu, oi, at $30°$

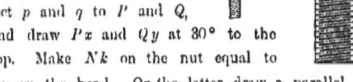
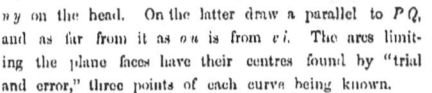

to the top, and cutting off the desired amount. Draw circles on the plans, with diameter equal to uo. Project p and q to P and Q, and draw Px and Qy at $30°$ to the top. Make Nk on the nut equal to ny on the head. On the latter draw a parallel to PQ, and as far from it as ou is from vi. The arcs limiting the plane faces have their centres found by "trial and error," three points of each curve being known.

When drawn to a small scale screws may be represented by either of the conventional methods illustrated by Figs. 314, 315 and 316.

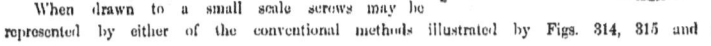

180   THEORETICAL AND PRACTICAL GRAPHICS.

DIMENSIONS OF BOLTS AND NUTS, UNITED STATES STANDARD (SELLERS SYSTEM)

Proportions of Bolt					Dimensions of Nuts Rough and Finished							Dimensions of Bolt Heads Rough and Finished					
Outside Diam. D	At Root of Thread		N = Number of Threads per inch.	Width (f) of Flat.	Across Flats		Across Corners		Across Corners		Depth of Nut		Across Flats		Across Corners		Across Corners	Depth of Head
	Diam.	Area			R	F	R	F	R	F	R	F	R	F	R	F	R	F
$\frac{1}{4}$	.185	.026	20	.0062	$\frac{1}{2}$	$\frac{7}{16}$	$\frac{37}{64}$		$\frac{7}{16}$		$\frac{1}{4}$	$\frac{3}{16}$	$\frac{1}{2}$	$\frac{7}{16}$	$\frac{37}{64}$		$\frac{7}{16}$	$\frac{3}{16}$
$\frac{5}{16}$	.240	.045	18	.0074	$\frac{19}{32}$	$\frac{17}{32}$	$\frac{11}{16}$		$\frac{5}{8}$		$\frac{5}{16}$	$\frac{1}{4}$	$\frac{19}{32}$	$\frac{17}{32}$	$\frac{11}{16}$		$\frac{5}{8}$	$\frac{1}{4}$
$\frac{3}{8}$	.294	.067	16	.0078	$\frac{11}{16}$	$\frac{5}{8}$	$\frac{51}{64}$		$\frac{23}{32}$		$\frac{3}{8}$	$\frac{5}{16}$	$\frac{11}{16}$	$\frac{5}{8}$	$\frac{51}{64}$		$\frac{23}{32}$	$\frac{5}{16}$
$\frac{7}{16}$	.344	.092	14	.0089	$\frac{25}{32}$	$\frac{21}{32}$	$\frac{9}{10}$		$1\frac{7}{64}$		$\frac{7}{16}$	$\frac{3}{8}$	$\frac{25}{32}$	$\frac{23}{32}$	$\frac{9}{10}$		$1\frac{1}{64}$	$\frac{3}{8}$
$\frac{1}{2}$	.400	.125	13	.0096	$\frac{7}{8}$	$\frac{13}{16}$	1		$1\frac{19}{64}$		$\frac{1}{2}$	$\frac{7}{16}$	$\frac{7}{8}$	$\frac{13}{16}$	1		$1\frac{15}{64}$	$\frac{7}{16}$
$\frac{9}{16}$	.454	.161	12	.0104	$\frac{31}{32}$	$\frac{29}{32}$	$1\frac{1}{8}$		$1\frac{21}{64}$		$\frac{9}{16}$	$\frac{1}{2}$	$\frac{31}{32}$	$\frac{29}{32}$	$1\frac{1}{8}$		$1\frac{21}{64}$	$\frac{1}{2}$
$\frac{5}{8}$	.507	.201	11	.0113	$1\frac{1}{16}$	1	$1\frac{7}{32}$		$1\frac{1}{2}$		$\frac{5}{8}$	$\frac{9}{16}$	$1\frac{1}{16}$	1	$1\frac{7}{32}$		$1\frac{1}{2}$	$\frac{9}{16}$
$\frac{3}{4}$	.620	.301	10	.0125	$1\frac{1}{4}$	$1\frac{3}{16}$	$1\frac{7}{16}$		$1\frac{40}{64}$		$\frac{3}{4}$	$\frac{11}{16}$	$1\frac{1}{4}$	$1\frac{3}{16}$	$1\frac{7}{16}$		$1\frac{40}{64}$	$\frac{11}{16}$
$\frac{7}{8}$	.731	.419	9	.0138	$1\frac{7}{16}$	$1\frac{3}{8}$	$1\frac{21}{32}$		$2\frac{1}{32}$		$\frac{7}{8}$	$\frac{13}{16}$	$1\frac{7}{16}$	$1\frac{3}{8}$	$1\frac{21}{32}$		$2\frac{1}{32}$	$\frac{13}{16}$
1	.837	.55	8	.0156	$1\frac{5}{8}$	$1\frac{9}{16}$	$1\frac{7}{8}$		$2\frac{19}{64}$		1	$\frac{15}{16}$	$1\frac{5}{8}$	$1\frac{9}{16}$	$1\frac{7}{8}$		$2\frac{19}{64}$	$\frac{15}{16}$
$1\frac{1}{8}$	.940	.693	7	.0178	$1\frac{13}{16}$	$1\frac{3}{4}$	$2\frac{3}{32}$		$2\frac{9}{16}$		$1\frac{1}{8}$	$1\frac{1}{16}$	$1\frac{13}{16}$	$1\frac{3}{4}$	$2\frac{3}{32}$		$2\frac{9}{16}$	$1\frac{1}{16}$
$1\frac{1}{4}$	1.065	.89	7	.0178	2	$1\frac{15}{16}$	$2\frac{1}{4}$		$2\frac{53}{64}$		$1\frac{1}{4}$	$1\frac{3}{16}$	2	$1\frac{15}{16}$	$2\frac{7}{16}$		$2\frac{49}{64}$	1	$1\frac{3}{16}$
$1\frac{3}{8}$	1.160	1.056	6	.0208	$2\frac{3}{16}$	$2\frac{1}{8}$	$2\frac{17}{32}$		$3\frac{5}{32}$		$1\frac{3}{8}$	$1\frac{5}{16}$	$2\frac{3}{16}$	$2\frac{1}{8}$	$2\frac{17}{32}$		$3\frac{5}{32}$	$1\frac{5}{16}$
$1\frac{1}{2}$	1.284	1.294	6	.0208	$2\frac{3}{8}$	$2\frac{5}{16}$	$2\frac{3}{4}$		$3\frac{23}{64}$		$1\frac{1}{2}$	$1\frac{7}{16}$	$2\frac{3}{8}$	$2\frac{5}{16}$	$2\frac{3}{4}$		$3\frac{23}{64}$	$1\frac{7}{16}$
$1\frac{5}{8}$	1.389	1.515	$5\frac{1}{2}$	.0227	$2\frac{9}{16}$	$2\frac{1}{2}$	$2\frac{31}{32}$		$3\frac{5}{8}$		$1\frac{5}{8}$	$1\frac{9}{16}$	$2\frac{9}{16}$	$2\frac{1}{2}$	$2\frac{31}{32}$		$3\frac{5}{8}$	$1\frac{9}{16}$
$1\frac{3}{4}$	1.491	1.746	5	.0250	$2\frac{3}{4}$	$2\frac{11}{16}$	$3\frac{3}{16}$		$3\frac{37}{64}$		$1\frac{3}{4}$	$1\frac{11}{16}$	$2\frac{3}{4}$	$2\frac{11}{16}$	$3\frac{3}{16}$		$3\frac{37}{64}$	$1\frac{3}{16}$	$1\frac{11}{16}$
$1\frac{7}{8}$	1.616	2.051	5	.0250	$2\frac{15}{16}$	$2\frac{7}{8}$	$3\frac{13}{32}$		$4\frac{5}{32}$		$1\frac{7}{8}$	$1\frac{13}{16}$	$2\frac{15}{16}$	$2\frac{7}{8}$	$3\frac{13}{32}$		$4\frac{5}{32}$	$1\frac{13}{16}$
2	1.712	3.301	$4\frac{1}{2}$	.0277	$3\frac{1}{8}$	$3\frac{1}{16}$	$3\frac{5}{8}$		$4\frac{27}{64}$		2	$1\frac{15}{16}$	$3\frac{1}{8}$	$3\frac{1}{16}$	$3\frac{5}{8}$		$4\frac{27}{64}$	$1\frac{15}{16}$
$2\frac{1}{8}$	1.962	3.023	$4\frac{1}{2}$	.0277	$3\frac{5}{16}$	$3\frac{1}{4}$	$4\frac{1}{3}$		$4\frac{51}{64}$		$2\frac{1}{8}$	$2\frac{3}{16}$	$3\frac{5}{16}$	$3\frac{1}{4}$	$4\frac{1}{3}$		$4\frac{51}{64}$	$1\frac{7}{8}$	$2\frac{3}{16}$
$2\frac{1}{4}$	2.176	3.718	4	.0312	$3\frac{1}{2}$	$3\frac{13}{16}$	$4\frac{1}{16}$		$5\frac{21}{64}$		$2\frac{1}{4}$	$2\frac{3}{16}$	$3\frac{1}{2}$	$3\frac{13}{16}$	$4\frac{1}{16}$		$5\frac{21}{64}$	$1\frac{15}{16}$	$2\frac{7}{16}$
$2\frac{3}{8}$	2.426	4.622	4	.0312	$4\frac{1}{4}$	$4\frac{3}{16}$	$4\frac{29}{32}$		6		$2\frac{3}{8}$	$2\frac{11}{16}$	4	$4\frac{3}{16}$	$4\frac{29}{32}$		6	$2\frac{1}{8}$	$2\frac{11}{16}$
3	2.629	5.428	$3\frac{1}{2}$	.0357	$4\frac{3}{8}$	$4\frac{5}{16}$	$5\frac{3}{8}$		$6\frac{17}{32}$		3	$2\frac{15}{16}$	$4\frac{3}{8}$	$4\frac{5}{16}$	$5\frac{3}{8}$		$6\frac{17}{32}$	$2\frac{5}{8}$	$2\frac{15}{16}$
$3\frac{1}{4}$	2.879	6.509	$3\frac{1}{2}$	.0357	5	$4\frac{15}{16}$	$5\frac{19}{32}$		$7\frac{7}{16}$		$3\frac{1}{4}$	$3\frac{3}{16}$	5	$4\frac{15}{16}$	$5\frac{19}{32}$		$7\frac{7}{16}$	$2\frac{3}{8}$	$3\frac{3}{16}$
$3\frac{1}{2}$	3.100	7.547	$3\frac{1}{4}$	.0384	$5\frac{3}{8}$	$5\frac{5}{16}$	$6\frac{5}{64}$		$7\frac{29}{64}$		$3\frac{1}{2}$	$3\frac{7}{16}$	$5\frac{3}{8}$	$5\frac{5}{16}$	$6\frac{5}{64}$		$7\frac{29}{64}$	$2\frac{11}{16}$	$3\frac{7}{16}$
$3\frac{3}{4}$	3.317	8.641	3	.0413	$5\frac{3}{4}$	$5\frac{11}{16}$	$6\frac{41}{32}$		$8\frac{1}{8}$		$3\frac{3}{4}$	$3\frac{11}{16}$	$5\frac{3}{4}$	$5\frac{11}{16}$	$6\frac{21}{32}$		$8\frac{1}{8}$	$2\frac{7}{8}$	$3\frac{11}{16}$
4	3.567	9.993	3	.0413	$6\frac{1}{8}$	$6\frac{1}{16}$	$7\frac{3}{32}$		$8\frac{41}{64}$		4	$3\frac{15}{16}$	$6\frac{1}{8}$	$6\frac{1}{16}$	$7\frac{3}{32}$		$8\frac{41}{64}$	$3\frac{1}{8}$	$3\frac{15}{16}$
$4\frac{1}{4}$	3.798	11.329	$2\frac{7}{8}$	.0435	$6\frac{1}{2}$	$6\frac{7}{16}$	$7\frac{1}{2}$		$9\frac{3}{16}$		$4\frac{1}{4}$	$4\frac{3}{16}$	$6\frac{1}{2}$	$6\frac{7}{16}$	$7\frac{1}{2}$		$9\frac{3}{16}$	$3\frac{1}{4}$	$4\frac{3}{16}$
$4\frac{1}{2}$	4.028	12.742	$2\frac{3}{4}$	.0454	$6\frac{7}{8}$	$6\frac{13}{16}$	$7\frac{29}{32}$		$9\frac{5}{8}$		$4\frac{1}{2}$	$4\frac{7}{16}$	$6\frac{7}{8}$	$6\frac{13}{16}$	$7\frac{29}{32}$		$9\frac{21}{32}$	$3\frac{3}{8}$	$4\frac{7}{16}$
$4\frac{3}{4}$	4.256	14.226	$2\frac{5}{8}$	.0476	$7\frac{1}{4}$	$7\frac{3}{16}$	$8\frac{13}{32}$		$10\frac{1}{4}$		$4\frac{3}{4}$	$4\frac{11}{16}$	$7\frac{1}{4}$	$7\frac{3}{16}$	$8\frac{13}{32}$		$10\frac{1}{4}$	$3\frac{1}{2}$	$4\frac{11}{16}$
5	4.480	15.763	$2\frac{1}{2}$	.0500	$7\frac{5}{8}$	$7\frac{9}{16}$	$8\frac{27}{32}$		$10\frac{49}{64}$		5	$4\frac{15}{16}$	$7\frac{5}{8}$	$7\frac{9}{16}$	$8\frac{27}{32}$		$10\frac{49}{64}$	$3\frac{5}{8}$	$4\frac{15}{16}$
$5\frac{1}{4}$	4.730	17.570	$2\frac{1}{2}$	.0500	8	$7\frac{15}{16}$	$9\frac{9}{32}$		$11\frac{23}{64}$		$5\frac{1}{4}$	$5\frac{3}{16}$	8	$7\frac{15}{16}$	$9\frac{9}{32}$		$11\frac{23}{64}$	4	$5\frac{3}{16}$
$5\frac{1}{2}$	4.953	19.267	$2\frac{3}{8}$	.0526	$8\frac{3}{8}$	$8\frac{5}{16}$	$9\frac{23}{32}$		$11\frac{7}{8}$		$5\frac{1}{2}$	$5\frac{7}{16}$	$8\frac{3}{8}$	$8\frac{5}{16}$	$9\frac{23}{32}$		$11\frac{7}{8}$	$4\frac{1}{8}$	$5\frac{7}{16}$
$5\frac{3}{4}$	5.203	21.261	$2\frac{3}{8}$	.0526	$8\frac{3}{4}$	$8\frac{11}{16}$	$10\frac{5}{32}$		$12\frac{3}{8}$		$5\frac{3}{4}$	$5\frac{11}{16}$	$8\frac{3}{4}$	$8\frac{11}{16}$	$10\frac{5}{32}$		$12\frac{3}{8}$	$4\frac{3}{8}$	$5\frac{11}{16}$
6	5.243	23.097	$2\frac{1}{4}$	.0555	$9\frac{1}{8}$	$9\frac{1}{16}$	$10\frac{19}{32}$		$12\frac{15}{16}$		6	$5\frac{15}{16}$	9	9	$10\frac{19}{32}$		$12\frac{15}{16}$	$4\frac{3}{8}$	$5\frac{11}{16}$

ORTHOGRAPHIC PROJECTION UPON A SINGLE PLANE.

CHAPTER XV.

AXONOMETRIC (INCLUDING ISOMETRIC) PROJECTION.—ONE-PLANE DESCRIPTIVE GEOMETRY.

621. When but one plane of projection is employed there are but two applications of orthographic projection having special names. These are *Axonometric* (known also as *Axometric*) *Projection*, and *One-Plane Descriptive Geometry* or *Horizontal Projection*.

AXONOMETRIC PROJECTION.— ISOMETRIC PROJECTION.

622. *Axonometric Projection*, including its much-employed special form of *Isometric Projection*, is applicable to the representation of the parts or "details" of machinery, bridges or other constructions in which the main lines are in directions that are mutually perpendicular to each other.

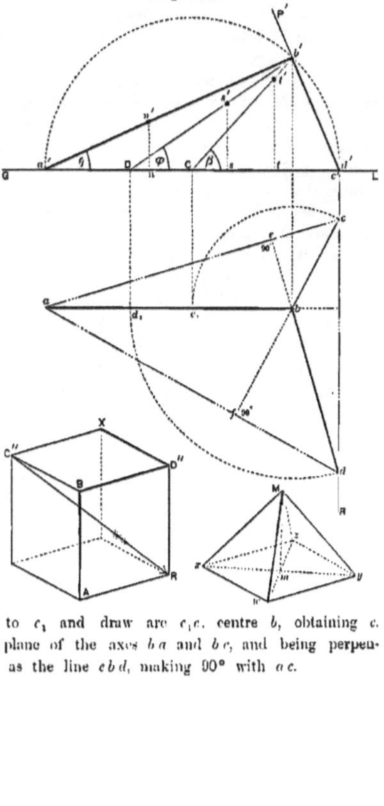

An axonometric drawing has a pictorial effect that is obtained with much less work than is involved in the construction of a true perspective, yet which answers almost as well for the conveying of a clear idea of what the object is; while it may also be made to serve the additional purpose of a working drawing, when occasion requires.

623. *Fundamental Problem.*— To obtain the orthographic projection of three mutually perpendicular lines or axes, and the scale of real to projected lengths. Let ab, bc and bd (Fig. 394) be the projections of three lines forming a solid right angle at b. Let the line ab be inclined at some given angle θ to the plane of projection. Locate a vertical plane parallel to ab and project the latter upon it at $a'b'_1$, at $\theta°$ to the horizontal. Since the plane of the other two axes is perpendicular to ab, $a'b'$, its traces will be $P'd'R$. (Art. 303).

In order to find either c or d we need to know the inclination of the axis having such point for its extremity. Supposing β given for cb, draw $b'C$ at $\beta°$ to G.L.; project C to c_1 and draw arc c_1c, centre b, obtaining c.

Join a with c; then ac is the trace of the plane of the axes ba and bc, and being perpendicular to the third axis we may draw the latter as the line ebd, making 90° with ac.

242   *THEORETICAL AND PRACTICAL GRAPHICS.*

Carry d to d_1 about b; project d_1 to D and join the latter with b'. Then Db' is the *true length*, and $b'DL$ (or ϕ) the *inclination*, of the third axis, bd.

Lay off $a'n$, Ds' and Ct', each one inch. Their projected lengths on the horizontal are respectively $a'n$, Ds and Ct. The latter are then the lengths, representative of inches, for all lines parallel to ab, bc and bd respectively.

624. *To make an axonometric projection of a one-inch cube*, to the scale just obtained.

Although not absolutely necessary *it is customary to take one axis vertical.*

Taking the ab-axis vertical, the cube in Fig. 394 fulfills the conditions. For BA equals $a'n$; BD'' equals Ds, and BC'' equals Ct, while the angles at B equal those at b.

The light being taken in the usual direction, i. e., parallel to the body-diagonal of the cube ($C''R$), the shade lines indicated are those which separate illumined from unillumined surfaces, and are those which could, therefore, cast shadows.

625. *The axonometric projection of a vertical pyramid*, of three-fourths-inch altitude and inch-square base, to the same scale as the cube. The pyramid in Fig. 394 meets the requirements, $xwyz$ having been made equal to $C''BD''X$; while the altitude mM, rising from the intersection of the diagonals of the base, equals three-fourths $a'n$, the inch-representative for the vertical axis.

626. *To draw curves in axonometric projection* obtain first the projections of their inscribed or circumscribed polygons, or of a sufficient number of secant lines; then sketch the curve through the points on these new lines which correspond to the points common to the curves and lines in the original figure. This will be illustrated fully in treating isometric projection.

627. *Isometric Projection.—Isometric Drawing.* When three mutually perpendicular axes are *equally inclined* to the plane of projection they will obviously make equal angles (120°) with each other in projection. This relation led to the name "isometric," implying equal measure, and also obviates the necessity for making a separate scale for each axis.

The advantages of this method seem to have been first brought out by Prof. Farish of England, who presented a paper upon it in 1820 before the Cambridge Philosophical Society of England.

628. In practice *the isometric scale is never used*, but, as all lines parallel to the axes are equally foreshortened, it is customary to lay off their given lengths directly upon the axes or their parallels, the result showing relative position and proportion of parts just as correctly as a true projection, but being then called an *isometric drawing*, to distinguish it from the other. It would, obviously, be the *projection* of a considerably larger object than that from which the dimensions were taken.

Lines parallel to the axes are called *isometric lines*.

Any plane parallel to, or containing two isometric axes, is called an *isometric plane*.

Fig. 395.   **629.** *To make an isometric drawing of a cube of three-quarter-inch edges.* Starting with the usual *isometric centre*, O, (Fig. 395) draw one axis vertical, and on it lay off OA equal to three-fourths of an inch. OC and OB are then drawn with the 30°-triangle as shown, made equal in length to OA, and the figure completed by parallels to the lines already drawn.

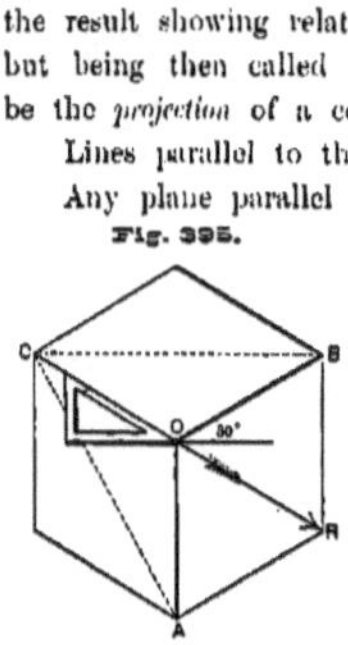

One body-diagonal of the cube is perpendicular to the paper at O.

630. *To draw circles and other curves isometrically*, employ auxiliary tangents and secants, obtain their isometric representations, and sketch the curves through the proper points.

In Fig. 396 we have an isometric cube, and at $MO'P'N$ the square, which—by rotation on MN and by an elongation of MP—becomes transformed into $MOPN$. The circle of centre S' then becomes the ellipse of centre S, whose points are obtained by means of the four tangencies d', F, E and G, and by making gn equal to gn', hm equal to $h'm'$, etc.

631. *The isometric circle may be divided* into parts corresponding to certain arcs on the original, either (1) by drawing radii from S' to MN, as those through b', c', d', (which may be equidistant or not, at pleasure) and getting their isometric representatives, which will intercept arcs, as bd', $d'c$, which are the isometric views of $b'd'$, $d'c'$; or (2) by drawing a semicircle xiy on the major axis as a diameter, letting fall perpendiculars to xy from various points, and noting the arcs as 1–2, 2–3, that are included between them and which correspond to the arcs ij, jk, originally assumed.

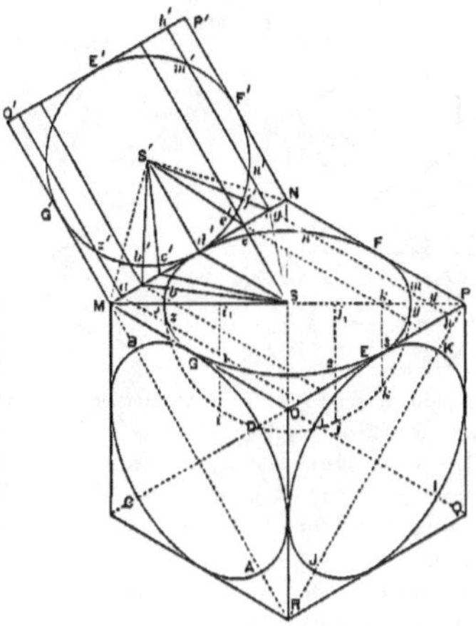

632. *Shade lines on isometric drawings.* While not universally adhered to, the conventional direction for the rays, in isometric shadow construction, is that of the body-diagonal CR of the cube (Fig. 395). This makes in projection an angle of 30° with the horizontal. Its projection on an isometrically-horizontal plane —as that of the top—is a horizontal line CB; while its projection CA, on the isometric representation of a vertical plane, is inclined 60° to the horizontal.

633. To illustrate the principles just stated Fig. 397 is given, in which all the lines are isometric, with the exception of Dz and its parallels, and ST. The drawing of non-isometric lines will be treated in the next article, but assuming the objects as given whose shadows we are about to construct, we may start with any line, as Dz.

The ray Dd is at 30°. Its projection d_1d is a horizontal through the plan of D. The ray and its projection meet at d. As the shadow begins where the line meets the plane, we have zd for the shadow of Dz. This gives the direction for the shadow of any line parallel to Dz, hence for yr, which, however, soon runs into the shadow of BC. As b is the intersection of the ray Bb with its projection b_1b, it is the shadow of B, and b_1b that of b_1B. Then br is parallel to BC, the line casting the shadow being parallel to the plane receiving it.

In accordance with the principle last stated, dc is equal and parallel to DE, and ef to EF. At f the shadow turns to g, as the ray jF, run back, cuts MG at j', and $j'G$ casts the jg-shadow.

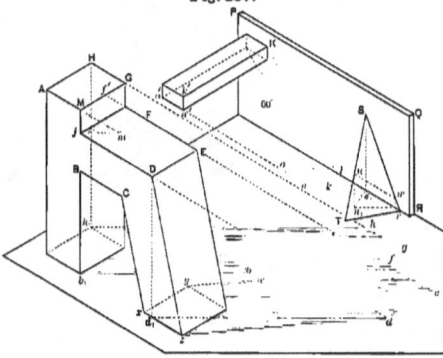

Fig. 396.

Fig. 397.

244   THEORETICAL AND PRACTICAL GRAPHICS.

Then yh equals GH, and hh_1 is the shadow of Hh. The projection jm catches the ray Mm at m. Then mf, equal to Mf', completes the construction.

The timber, projecting from the vertical plane PQR, illustrates the 60°-angle earlier mentioned. Kk' being perpendicular to the vertical plane, its shadow $K'l$ is at 60° to the horizontal, and $K'lk$ is the plane of rays containing said edge. Its horizontal trace catches the ray from k' at k. Then nk, the shadow of $n'k'$, is horizontal, being the trace of a vertical plane of rays on an isometrically-horizontal plane. The construction of the remainder is self-evident.

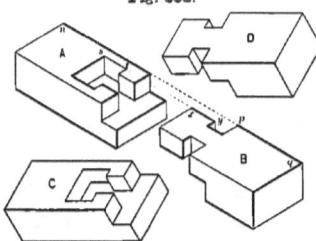

Fig. 398.

Letting ST represent a small rod, oblique to isometric planes, assume any point on it, as u; find its plan, u_1; take the ray through u and find its trace w. Then Sw is the direction of the shadow on the vertical plane, and at r it runs off the vertical and joins with T.

634. *Timber framings, drawn isometrically*, are illustrated by Figures 398 and 399. In Fig. 398 the pieces marked A and B show one form of mortise and tenon joint, and are drawn with the lines in the customary directions of isometric axes. The same pieces are represented again at C and D, all the lines having been turned through an angle of 30°, so that while maintaining the same relative direction to each other and being still truly isometric, they lie differently in relation to the edges of the paper—a matter of little importance when dealing with comparatively small figures, but affecting the appearance of a large drawing very materially.

635. *Non-isometric lines.—Angles in isometric planes.* In Fig. 399 a portion of a cathedral roof truss is drawn isometrically.

Three pieces are shown that are not parallel to isometric lines. To represent them correctly we need to know the real angles made by them with horizontal or vertical pieces, and use isometric coördinates or "offsets" in laying them out on the drawing.

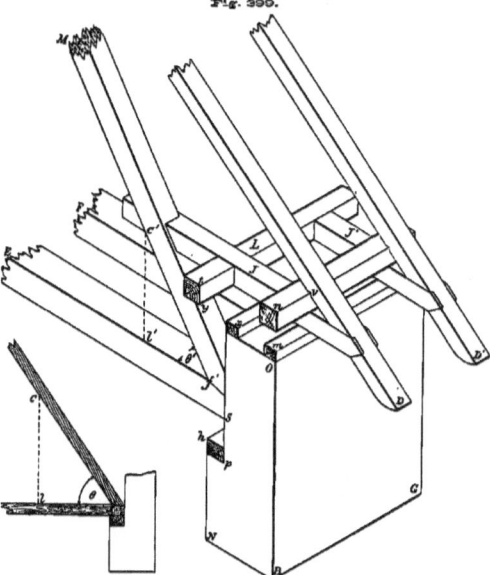

Fig. 399.

In the lower figure we see at θ the actual angle of the inclined piece Mf' to the horizontal. Offsets, fl and lC, to any point C of the inclined piece, are laid off in isometric directions at $f'l'$

and $l'C'$, when $C'f'l'$ (or $6'$) is the isometric view of 6. A similar construction, not shown, gave the directions of pieces D and D'.

Much depends on the choice of the *isometric centre*. Had N been selected instead of B, the top surfaces of the inclined pieces would have been nearly or quite projected in straight lines, rendering the drawing far less intelligible.

The student will notice that the shade lines on Fig. 399 are located *for effect*, and in violation of the usual rule, it having been found that the best appearance results from assuming the light in such direction as to make the most shade lines fall centrally on the timbers.

636. *Non-isometric lines.—Angles not in isometric planes.* To draw lines not lying in isometric planes requires the use of three isometric offsets. As one of the most frequent applications of isometric drawing is in problems in stone cutting, we may take one such to advantage in illustrating constructions of this kind.

Fig. 400 shows an arched passage-way, in plan and elevation. The surface $n\,o$, $r'l'n'o'$ is vertical as far as $n'o'$, and conical (with vertex J, C') from there to $n''o''$. The vertical surface on $n\,n$ is tangent at n' to the cylinder $n'f'e'o''$. Similarly, $m\,m$ is vertical to m', and there changes into the cylinder $m'y'h'$.

The radial bed $b'y'$ is indicated on the plan (though not in full size) by parallel lines at $bc'figzb$. The bed $a'h'$ is of the same form as $b'y'$, being symmetrical with it.

In Fig. 401 we have an enlarged drawing of the keystone with the plan inverted, so that all the faces of the stone may be correctly represented as seen. The isometric drawing is made to correspond, that is, it represents the stone after a 180°-rotation about an axis perpendicular to the paper.

The isometric block in which the keystone can be inscribed is shown in dotted lines, its dimensions, derived from the projections, being *length*, $AA_1 = aa$; *breadth*, $AB = a'b'$; *height*, $AO = a'p$.

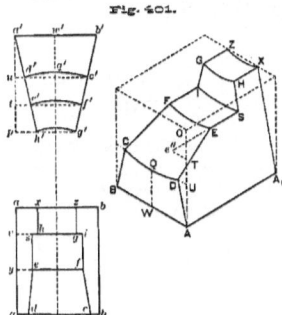

Fig. 401.

The top surface $a'b'$ becoming the lower in the isometric, reverses the direction of the lines. Thus, a' is seen at A, and b' at B. To get D make $AU = a'u$, then $UD = ud'$. Make C symmetrical with D and join with B, and also D with A. WQ equals $w'q'$, for the ordinate of the middle point of the arc.

DE is not an isometric plane, hence to reach E from A we make $AT = a't$; $Tc'' = tc'$, and $c''E = ay$ (the distance of c from the plane ab).

The remainder of the construction is but a duplication of one or other of the above processes.

The principle that lines that are parallel on the object will also be parallel on the drawing may be frequently availed of in the interest of rapid construction or for a check as to accuracy.

THEORETICAL AND PRACTICAL GRAPHICS.

HORIZONTAL PROJECTION OR ONE-PLANE DESCRIPTIVE GEOMETRY.

637. *One-Plane Descriptive Geometry* or *Horizontal Projection* is a method of using orthographic projections with but one plane, the fundamental principle being that the space-position of a point is known if we have its projection on a plane and also know its distance from the plane.

Thus, in Fig. 402, a with the subscript 7 shows that there is a point A, vertically above a and at seven units distance from it. The significance of b_3 is then evident, and to show the line in its true length and inclination we have merely to erect perpendiculars aA and Bb, of *seven* and *three* units respectively, join their extremities, and see the line AB in true length and inclination.

Fig. 402.

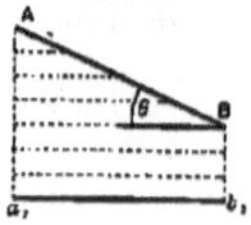

In this system the horizontal plane alone is used; One-plane Descriptive is therefore applied only to constructions in which the lines are mainly or entirely horizontal, as in the mapping of small topographical or hydrographical surveys, in which the curvature of the earth is neglected; also in drawing fortifications, canals, etc.

The plane of projection, usually called the *datum* or *reference* plane, is taken, ordinarily, below all the points that are to be projected, although when mapping the bed of a stream or other body of water it is generally taken at the water line, in which case the numbers, called *indices* or *references*, show *depths*.

638. A *horizontal* line evidently needs but one index. This is illustrated in mapping *contour lines*, which represent sections of the earth's surface by a series of equidistant horizontal planes.

Fig. 403.

HILL CONTOUR

In Fig. 403 the curves indicate such a series of sections made by planes one yard, metre or other unit apart, the larger curve being assumed to lie in the reference or datum plane, and therefore having the index *zero*.

The profile of a section made by any vertical plane MN would be found by laying off—to any assumed scale for vertical distances—ordinates from the points where the plane cuts the contours, giving each ordinate the same number of units as are in the index of the curve from which it starts. Such a section is shown in the shaded portion on the left, on a ground line PQ, which represents MN transferred.

Fig. 404.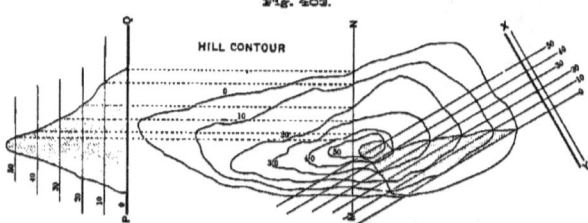

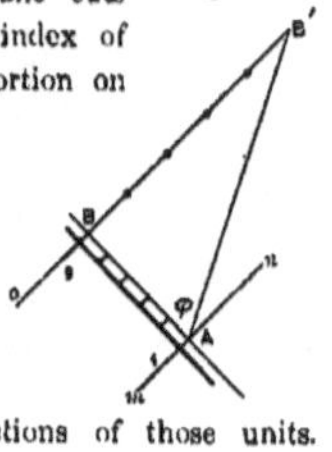

639. The steepness of a plane or surface is called its *slope* or *declivity*. A *line of slope* is the steepest that can be drawn on the surface. A *scale of slope* is obtained by graduating the plan of a line of slope so that each unit on the scale is the projection of the unit's length on the original line. Thus, in Fig. 404, if ma and oB are horizontal lines in a plane, one having the index 4 and the other 9, the point B is evidently five units above A, and the five equal divisions between it and A are the projections of those units.

HORIZONTAL PROJECTION OR ONE-PLANE DESCRIPTIVE. 247

The scale of slope is often used as a ground line upon which to get an edge view of the plane. Thus, if BB' is at $90°$ to BA, and its length five units, then $B'A$ is the plane, and ϕ is its inclination.

The scale of slope is always made with a double line, the heavier of the two being on the left, ascending the plane.

As no exhaustive treatment of this topic is proposed here, or, in fact, necessary, in view of the simplicity of most of the practical applications and the self-evident character of the solutions, only two or three typical problems are presented.

640. *To find the intersection of a line and plane.* Let $a_{15} b_{30}$ be the line, and XY the plane. Draw horizontal lines in the plane at the levels of the indexed points. These, through 15 and 30 on XY, meet horizontal lines through a and b at c and d; cd is then the line of intersection of XY and a plane containing ab; hence c is the intersection of the latter with XY.

Fig. 405.

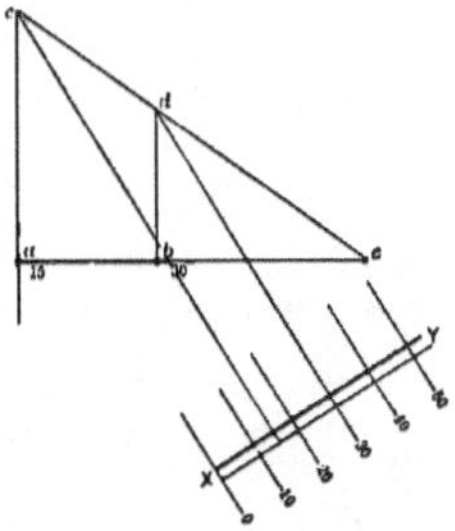

The same point c would have resulted if the lines ac and bd had been drawn in any other direction while still remaining parallel.

Fig. 406.

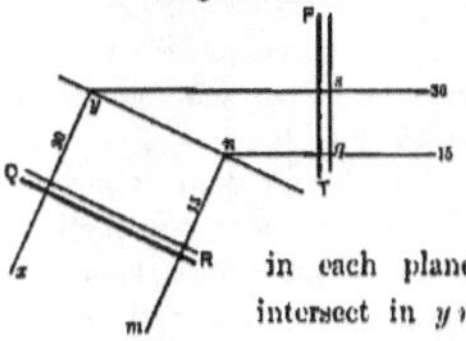

641. *To obtain the line of intersection of two planes,* draw two horizontals in each, at the same level, and join their points of intersection.

In Fig. 406 we have mn and qn as horizontals at level 15, one in each plane. Similarly, xy and ys are horizontals at level 30. The planes intersect in yn.

Were the scales of slope parallel, the planes would intersect in a horizontal line, one point of which could be found by introducing a third plane, oblique to the given planes, and getting its intersection with each, then noting where these lines of intersection met.

642. *To find the section of a hill by a plane of given slope.* Draw, as in the problem of Art. 640, horizontal lines in the plane, and find their intersections with contours at the same level. Thus, in Fig. 403, the plane XY cuts the hill in the shaded section nearest it, whose outlines pass through the points of intersection of horizontals 10, 20, 30 of the plane, with the like-numbered contour lines.

CHAPTER XVI.

OBLIQUE OR CLINOGRAPHIC PROJECTION.—CAVALIER PERSPECTIVE.—CABINET PROJECTION. MILITARY PERSPECTIVE.

643. If a figure be projected upon a plane by a system of parallel lines that are oblique to the plane, the resulting figure is called an *oblique* or *clinographic projection*, the latter term being more frequently employed in the applications of this method to crystallography. Shadows of objects in the sunlight are, practically, oblique projections.

In Fig. 407, $ABnm$ is a rectangle and $mxyn$ its oblique projection, the parallel projectors Ax and By being inclined to the plane of projection.

644. *When the projectors make $45°$ with the plane* this system is known either as *Cavalier Perspective*,

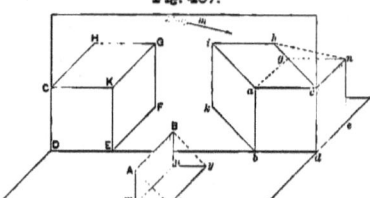
Fig. 407.

Cabinet Projection or *Military Perspective*, the plane of projection being vertical in the case of the first two, and horizontal in the last.

645. *Cavalier Perspective.—Cabinet Projection.— Military Perspective*. As just stated, the projectors being inclined at $45°$ for the system known by the three names above, we note that in this case a line perpendicular to the plane of projection, as Am or Bn (Fig. 407), will have a projection equal to itself. It is, therefore, unnecessary to draw the rays for lines so situated, as the known original lengths can be directly laid out on lines drawn in the assumed direction of projections.

Let $abcd.n$ be a cube with one face coinciding with the vertical plane. If the arrow m indicates one direction of rays making $45°$ with V, then the ray hn, parallel to m, will give h as the projection of n, and from what has preceded we should have ch equal to cn, and analogously for the remaining edges, giving $abcd.i$ for the cavalier perspective of the cube.

Similarly, EKH is a correct projection of the same cube for another direction of projectors, and we may evidently draw the oblique edges in any other direction and still have a cavalier perspective, by making the projected line equal to the original, when the latter is perpendicular to the plane of projection.

646. *Oblique projection of circles.* Were a circle inscribed in the back face of the cube DKG (Fig. 407) the projectors through the points of the circle would give an oblique *cylinder of rays*, whose intersection with the vertical plane DX would be a circle, since parallel planes cut a cylinder in similar sections. We see, therefore, that the oblique projection of a circle is itself circular when the plane of projection is parallel to that of the circle. In any other case the oblique projection of a circle may be found like an isometric projection (see Art. 631), viz., by drawing chords of the circle, and tangents, then representing such auxiliary lines in oblique view and sketching the curve (now an ellipse) through the proper points. Fig. 408 illustrates this in full.

OBLIQUE PROJECTION.—CAVALIER PERSPECTIVE.

647. *Oblique projection* is even better adapted than isometric to the representation of timber framings, machine and bridge details, and other objects in which straight lines—usually in mutually perpendicular directions—predominate, since all angles, curves, etc., lying in planes parallel to the paper, appear of the same *form* in projection, while the relations of lines perpendicular to the paper are preserved by a simple ratio, ordinarily one of equality.

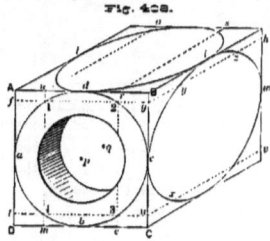

Fig. 408.

648. When the rays make with the plane of projection an angle greater than 45°, oblique projections give effects more closely analogous to a true perspective, since the foreshortening is a closer approximation to that ordinarily existing from a finite point of view. This is illustrated by Fig. 409, in which an object $ABDE$, known to be 1" thick, has its depth represented as only ½" in the second view, instead of full size, as in a cavalier perspective, the front faces being the same size in each. Provided that the scale of reduction were known, $abcdkf$ would answer as well for a working drawing as a 45°-projection.

649. By way of contrast with an isometric view the timber framing represented by Fig. 398 is

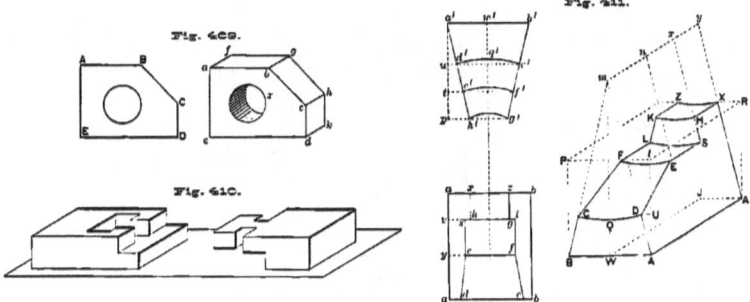

Fig. 409.

Fig. 410.

Fig. 411.

drawn in cavalier perspective in Fig. 410. Reference may advantageously be made, at this point, to Figs. 44, 45 and 46, which are oblique views of one form.

The keystone of the arch in Fig. 400, whose isometric view is shown in Fig. 401, appears in oblique projection in Fig. 411; the direction of lines not parallel to the axes of the circumscribing prism being found by "offsets" that must be taken in axial directions.

Fig. 412.

650. *Shadows, in oblique projection.* As in other projections, the conventional direction for the light is that of the body-diagonal of the oblique cube. The edges to draw in shade lines are obvious on inspection. (Fig. 412.)

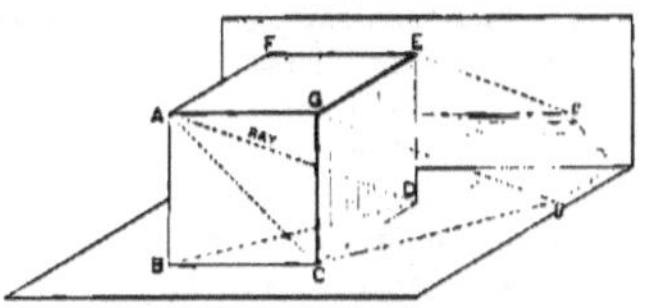

651. An interesting application of oblique projection, earlier mentioned, is in the drawing of crystals. Fig. 414 illustrates this, in the representation of a form common in fluorite and called the tetrahexahedron, bounded by twenty-four planes, each of which fulfills the condition expressed in the formula

250   THEORETICAL AND PRACTICAL GRAPHICS.

$\infty : n : 1$; that is, each face is parallel to one axis, cuts another at a unit's distance, and the third at some multiple of the unit.

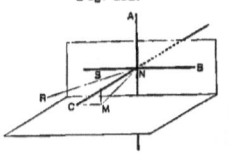
Fig. 413.

The three axes in this system are equal, and mutually perpendicular; but their projected lengths are aa', bb', cc'.

The direction of projectors which was assumed to give the lengths shown, was that of RN in Fig. 413, derived by turning the perpendicular CN through a horizontal angle $CNM = 18°\,26'$, and then elevating it through a vertical angle $MNS = 9°\,28'$; values that are given by Dana as well adapted to the exhibition of the forms occurring in this system.

The axes once established, if we wish to construct on them the form $\infty : 2 : 1$, we lay off on

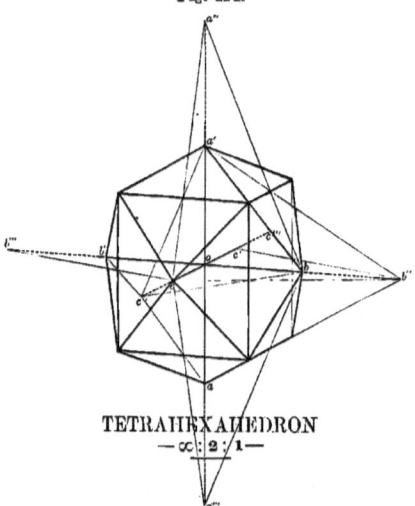
Fig. 414.

TETRAHEXAHEDRON
— $\infty : 2 : 1$ —

each (extended) one-half its own (projected) length; thus cc'' and $c'c'''$ each equal oc'; bb'' equals ob, etc. Then draw in light lines the traces of the various faces on the planes of the axes. Thus, $a'b''$ and $a''b$ each represent the trace of a plane cutting the c-axis at infinity, and the other axes at either one or two units distance; the former intercepting the *two* units on the b-axis and the *one* on the a-axis, while for $a''b$ it is exactly the reverse. Through the intersection of $a'b''$ and $a''b'$ a line is drawn parallel to the c-axis, indefinite in length at first, but determinate later by the intersection with it of other edges similarly found.

The student may develop in the same manner the forms $\infty : 3 : 1$; $\infty : 2 : 3$; $\infty : 3 : 4$; $\infty : 4 : 5$.

APPENDIX

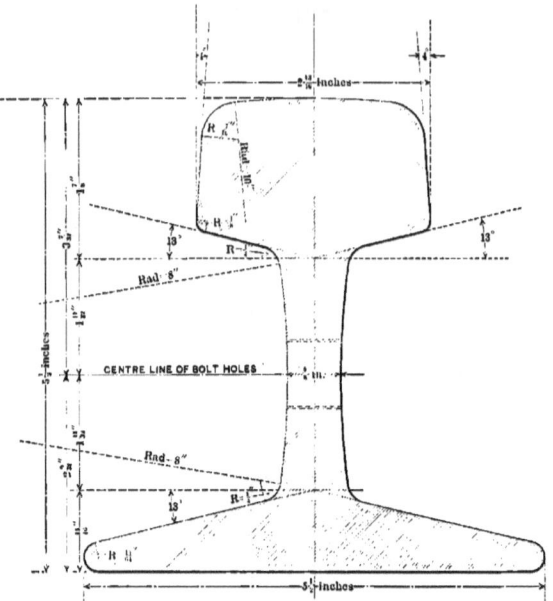

P. R. R. STANDARD RAIL SECTION.

100 LBS. PER YARD.

Draw the above either full size or enlarged 50%. In either case draw section lines in Prussian blue, spacing not less than one-twentieth of an inch. Dimension lines, red. Dimensions and arrow heads, black. Lettering and numerals either in Extended Gothic or Reinhardt Gothic.

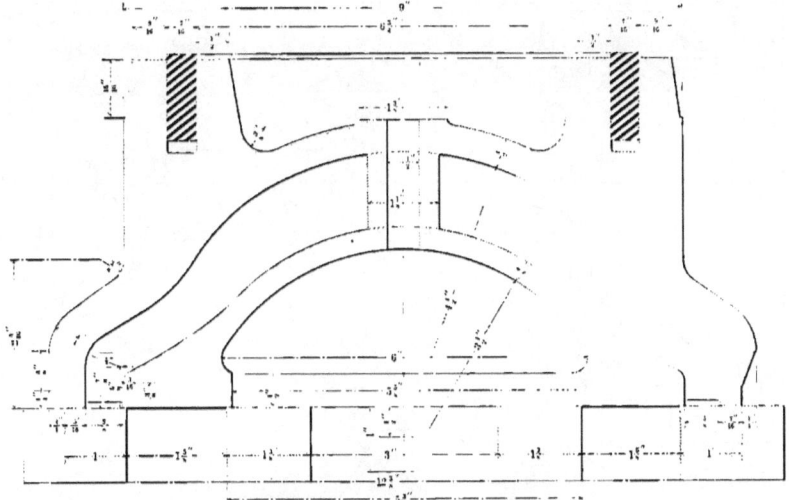

ALLEN-RICHARDSON SLIDE VALVE.

Draw either full size or larger. Section lines in Prussian blue, one-twentieth of an inch apart. Dimension lines, red. Dimensions and arrow heads, black. Lettering and numerals either in Extended Gothic or Reinhardt Gothic.

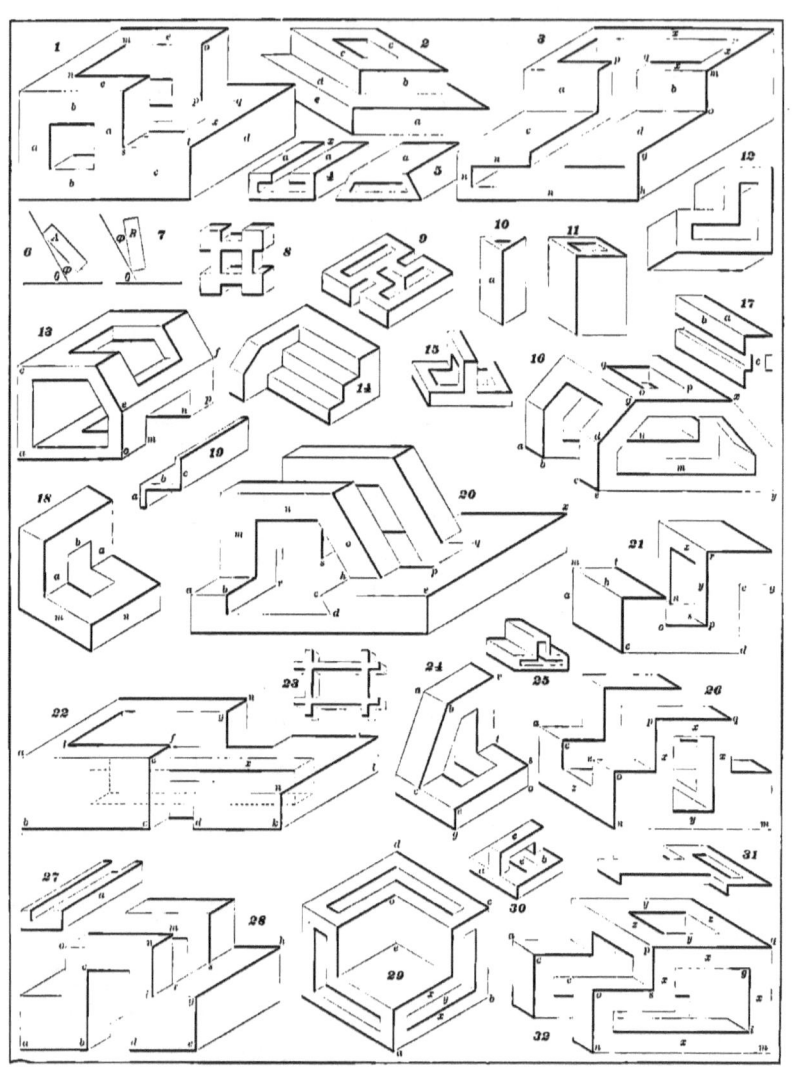

No. 1.

A B C D E F G H I J K L M N O P Q R S T U V W X Y Z &
a b c d e f g h i j k l m n o p q r s t u v w x y z 1 2 3 4 5 6 7 8 9 0

No. 2.

A B C D E F G H I J K L M N O P Q R S T
U V W X Y Z & 1 2 3 4 5 6 7 8 9 0 , .

No. 3.

A B C D E F G H I J K L M N O P Q R S T U V W X Y Z &
1 2 3 4 5 6 7 8 9 0

No. 4.

A B C D E F G H I J K L M N O P Q R S T U V W X Y Z & , .

No. 5.

A B C D E F G H I J K L M N O P Q R S T U V W X Y Z &
1 2 3 4 5 6 7 8 9 0

No. 6.

A B C D E F G H I J K L M N O P Q R S T U V W X Y Z &
1 2 3 4 5 6 7 8 9 0

No. 7.

A B C D E F G H I J K L M N O P Q R S T U V W X Y Z &
1 2 3 4 5 6 7 8 9 0

No. 8.

A B C D E F G H I J K L M N O P Q R S T U V W X Y Z &
a b c d e f g h i j k l m n o p q r s t u v w x y z . . 1 2 3 4 5 6 7 8 9 0

No. 9.

*A B C D E F G H I J K L M N O P Q R S T U V W
X Y Z & a b c d e f g h i j k l m n o p q r s t u v w x y z .,
1 2 3 4 5 6 7 8 9 0*

No. 10.

ABCDEFGHIJKLMNOPQRSTUVWX
YZ&abcdefghijklmnopqrstuvw
xyz.,1234567890

No. 11.

ABCDEFGHIJKLMNOPQRS
TUVWXYZ&
abcdefghijklmnopqrstuvwxyz
1234567890

No. 12.

ABCDEFGHIJKLMNOPQRSTUVWXYZ&
abcdefghijklmnopqrstuvwxyz.,1234567890

No. 13.

ABCDEFGHIJKLMNOPQRST
UVWXYZ&abcdefghijklmno
pqrstuvwxyz1234567890&

No. 14.

ABCDEFGHIJKLMNOPQRST
UVWXYZ&abcdefghijklmnop
qrstuvwxyz1234567890

No. 15.

A B C D E F G H I J K L M N O P Q R S T
U V W X Y Z & a b c d e f g h i j k l m n o
p q r s t u v w x y z 1 2 3 4 5 6 7 8 9 0

No. 16.

A B C D E F G H I J K L M N O P Q R S T U V
W X Y Z & a b c d e f g h i j k l m n o p q r s t
u v w x y z 1 2 3 4 5 6 7 8 9 0

No. 17.

𝔄 𝔅 ℭ 𝔇 𝔈 𝔉 𝔊 ℌ 𝔍 𝔍 𝔎 𝔏 𝔐 𝔑 𝔒 𝔓 𝔔 ℜ 𝔖
𝔗 𝔘 𝔙 𝔚 𝔛 𝔜 𝔞 𝔟 𝔠 𝔡 𝔢 𝔣 𝔤 𝔥 𝔦 𝔧 𝔨 𝔩
𝔪 𝔫 𝔬 𝔭 𝔮 𝔯 𝔰 𝔱 𝔲 𝔳 𝔴 𝔵 𝔶 𝔷 1 2 3 4 5 6 7 8 9 0

No. 18.

A B C D E F G H I J K L M N O P Q R S T U V W X Y Z &
a b c d e f g h i j k l m n o p q r s t u v w x y z . , 1 2 3 4 5 6 7 8 9 0

No. 19.

A B C D E F G H I J K L M N O P Q R S T U
V W X Y Z & a b c d e f g h i j k l m n o p q r s
t u v w x y z 1 2 3 4 5 6 7 8 9 0

No. 20.

A B C D E F G H I J K L M N O
P Q R S T U V W X Y Z & a b c
d e f g h i j k l m n o p q r s t u
v w x y z 1 2 3 4 5 6 7 8 9 0

No. 21.

A B C D E F G H I J K L M N O P Q R S T U V W X Y Z &
1 2 3 4 5 6 7 8 9 0 , .

No. 22.

ABCDEFGHIJKLMNOPQRSTUVWXYZ&1234567890

No. 23.

*A B C D E F G H I J K L M N O P Q R S
T U V W X Y Z & a b c d e f g h i j k l m n
o p q r s t u v w x y z . , 1 2 3 4 5 6 7 8 9 0*

No. 24.

A B C D E F G H I J K L M N O P Q R S T U V W X Y Z &
a b c d e f g h i j k l m n o p q r s t u v w x y z 1 2 3 4 5 6 7 8 9 0

No. 25.

A B C D E F G H I J K L M N O P Q R S T U V W X Y Z
& 1 2 3 4 5 6 7 8 9 0 , .

No. 26.

A B C D E F G H I J K L M N O P Q R S T U V
W X Y Z & A B C D E F G H I J K L M N O P Q R S T
U V W X Y Z . , 1 2 3 4 5 6 7 8 9 0

No. 27.

A B C D E F G H I J K L M N
O P Q R S T U V W X Y Z &
1 2 3 4 5 6 7 8 9 0 , .

No. 28.

ABCDEFGHIJKLMNOPQRSTUVW
XYZ&abcdefghijklmnopqrstuvw
xyz.,1234567890

No. 29.

ABCDEFGHIJKLMNOPQRST
UVWXYZ&abcdefghijklmn
opqrstuvwxyz.,1234567890

No. 30.

ABCDEFGHIJKLMNOPQRSTUV
WXYZ&1234567890:;..

No. 31.

ABCDEFGHIJKLMNO
PQRSTUVWXYZ&abcd
efghijklmnopqrstuvwxyz
1234567890

No. 32.

ABCDEFGHIJKLMNOPQRS
TUVWXYZ&.,1234567890

No. 33.

ABCDEFGHIJKLMNOPQRS
TUVWXYZ&abcdefghijklmn
opqrstuvwxyz1234567890 ⊙ ※ ∴

No. 34.

ABCDEFGHIJKLMNOPQRS
TUVWXYZ&abcdefghijklmn
opqrstuvwxyz1234567890

No. 35.

ABCDEFGHIJKL
MNOPQRSTUVW
XYZ&abcdefghijkl
mnopqrstuvwxyz.,
1234567890

No. 36.

ABCDEFGHIJKLMNOPQRS
TUVWXYZ&1234567890

No. 37.

ABCDEFGHIJKLMNOPQRS
TUVWXYZ&abcdefghijklm
nopqrstuvwxyz.,1234567890

No. 38.

AAaBBCDDDEeFGHHh
iJJKLMNNoOPQRSTU
VWXYZ&1234567890

No. 39.

A B C D E F G H I J K L M N O P
Q R S T U V W X Y Z & a b c d e f
g h i j k l m n o p q r s t u v w x y z
1 2 3 4 5 6 7 8 9 0

No. 40.

A B C D E F G H I J K L M N O P
Q R S T U V W X Y Z & a b c d e f
g h i j k l m n o p q r s t u v w x y z
1 2 3 4 5 6 7 8 9 0

No. 41.

A B C D E F G H I J K L M N O P Q R
S T U V W X Y Z &
a b c d e f g h i j k l m n o p q r s t u v w
x y z 1 2 3 4 5 6 7 8 9 0

No. 42.

A B C D E F G H I J K L M N O P
Q R S T U V W X Y Z &
❊ 1 2 3 4 5 ✺ 6 7 8 9 0 ❊

No. 43.

A B C D E F G H I J K L M N O P Q R S T U
V W X Y Z & a b c d e f g h i j k l m n o p q r
s t u v w x y z 1 2 3 4 5 6 7 8 9 0

No. 44.

𝔄𝔅ℭ𝔇𝔈𝔉𝔊ℌℑ𝔍𝔎𝔏𝔐𝔑𝔒𝔓𝔔ℜ𝔖
𝔗𝔘𝔙𝔚𝔛𝔜ℨ & a b c d e f g h i j k l m n o p q r
s t u v w x y z , .

No. 45.

A B C D E F G H I J K L M N O P
Q R S T U V W X Y Z &
1 2 3 4 5 6 7 8 9 0

No. 46.

𝔄𝔅ℭ𝔇𝔈𝔉𝔊ℌℑ𝔍𝔎𝔏𝔐𝔑𝔒
𝔓𝔔ℜ𝔖𝔗𝔘𝔙𝔚𝔛𝔜ℨ & a b c d
e f g h i j k l m n o p q r s t u v w x y z . ,
1 2 3 4 5 6 7 8 9 0

No. 47.

A B C D E F G H I J K L
M N O P Q R S T U
V W X Y Z &

No. 48.

A B C D E F G H I J K L M N
O P Q R S T U V W X Y Z &
1 2 3 4 5 6 7 8 9 0

No. 49.

A B C D E F G H I J K L M N O P
Q R S T U V W X Y Z &
← 1 2 3 4 5 ✢ 6 7 8 9 0 →

No. 50.

A B C D E F G H I J K L M N O P Q
R S T U V W X Y Z &
1 2 3 4 5 6 7 8 9 0

No. 51.

A B C D E F G H I J K L M N
O P Q R S T U V W X Y Z
1 2 3 4 5 6 7 8 9 0

No. 52.

ABCDEFGHIJKLMNOPQRSTU
VWXYZ&abcdefghijklmnopqr
stuvwxyz.,1234567890

No. 53.

ABCDEFGHIJKLMNOPQR
STUVWXYZ&1234567890,.

No. 54.

ABCDEFGHIJKLMNOPQR
STUVWXYZ&abcdefghijkl
mnopqrstuvwxyz1234567890

No. 55.

ABCDEFGHIJKLMN
OPQRSTUVWXYZ&
1234567890

No. 56.

ABCDEFGHIJKLMNOPQRSTUVW
XYZ&abcdefghijklmnopqrstuvw
xyz.,1234567890

No. 57.

ABCDEFGHIJK
LMNOPQRSTU
VWXYZ&12345
67890,.

No. 58.

ABCDEFGHIJKLMNOPQRS
TUVWXYZ&abcdefghijk
lmnopqrstuvwxyz
1234567890

No. 59.

ABCDEFGHIJKLMNO
PQRSTUVWXYZ&123
4567890,.

No. 60.

ABCDEFGHIJK
LMNOPQRSTUV
WXYZ

No. 61.

ABCDEFGHIJKLMNO
PQRSTUVWXYZ abcd
efghijklmnopqrstuvwxyz.,
1234567890

No. 62.

A B C D E F G
H I J K L M N
O P Q R S T U
V W X Y Z

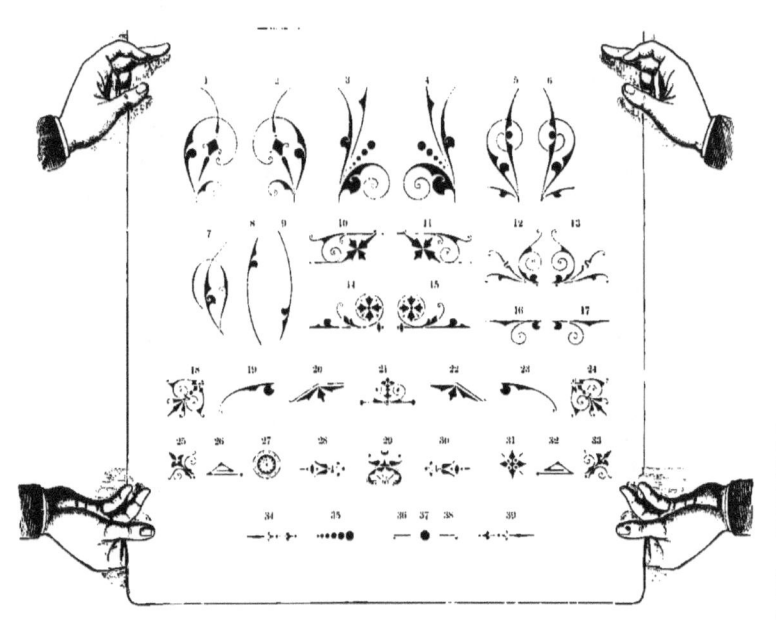

www.ingramcontent.com/pod-product-compliance
Lightning Source LLC
Chambersburg PA
CBHW032149160426
43197CB00008B/837